Understanding Non-Linear Strain-Displacement Relations In The Elliptical Co-ordinate System
(with applications to holes in a thin plate)

Rajen Merchant

Dedication

This book is dedicated to

my wife

Rekha

Preface

This second monograph is also specially written for students and researchers in mechanical and/or civil engineering, solid mechanics, materials science, physics, and mathematics. Professors teaching this subject will find it very useful.

The first monograph (titled ***Understanding the Elastic Stress Field Around an Elliptical Hole in a Thin Plate (in the deformed configuration)- ISBN 978-1-48359-262-6***) utilized stress equilibrium conditions to develop expressions for stress components which contain a parameter α_e. This second monograph does not deal with stress equilibrium conditions. It deals with strain-displacement relations as well as stress-strain relations and shows how to evaluate α_e for linear elastic materials.

I thank Prof. Rajendra Dubey for introducing me to the basic concepts of current configuration and the use of Elliptical co-ordinate system. Prof. Mahesh Pandey provided initial financial support for Phase 1 of this research. Prof. G. Glinka often set aside time out of his busy schedule whenever I requested help. I am indeed thankful to Prof. Katerina Papoulia for providing partial financial support for Phase 1 of this research as well as pointing out difficulties in the original model. However, most of this Phase 2 research was developed after 2013. Dr. Amin Eshraghi devoted (and continues to devote) a lot of time verifying my work. Dr. Anup Sahoo also spent a considerable amount of time reviewing my work and providing me encouragement when I needed it most. His guidance was useful since he had worked on similar models.

The University of Waterloo provided some financial support while I was registered as a graduate student. Prof. Rashmi Desai was always available for any difficulties I faced. I am thankful to Prof. Amit Singh for a detailed review of this second book and Prof. Jit Sharma for writing the Foreword. Prof. Bhalchandra Puranik's guidance was very helpful.

I sincerely thank Rekha (my wife) for logistical support and being supportive when I was spending hours at a time analyzing various scenarios. My children Reena and Himanshu helped a lot in the production of this monograph and related video presentations. My special thanks to the publisher bookbaby.com for taking care of all details related to printing and distribution of this monograph.

I bear full responsibility for the content and presentation including any errors and omissions in this work. However, I and the publisher assume no responsibility for damages resulting from the use of the information contained herein. I will greatly appreciate any comments and suggestions. I can be reached at 1-416-725-0909 or rajen.merchant@gmail.com.

I appeal to you not to copy any part of this monograph. Monographs are usually not sold in large numbers. Furthermore, monographs that involve complicated mathematical expressions are not easy to typeset. I have tried my best to keep the cost of each book to an affordable level so that as many students (and researchers) can buy, read, understand, and apply these concepts in their work.

Rajen Merchant

Toronto, August 2019

Foreword

It gives me great pleasure to write this foreword for Mr. Rajen Merchant's second book on the stress-strain analysis of an elliptical hole (representing a defect in a continuum) in an infinite thin plate. The first book focused exclusively on the evaluation of stress field based upon the stress equilibrium conditions. The Airy stress functions were developed in the first book so as to satisfy the field equations of equilibrium, expressed in the current (deformed) configuration. The results presented in the first book showed stress concentrations that are orders of magnitude higher in certain conditions once the defect changes shape. This second book incorporates non-linear strain-displacement relations as well as a constitutive (stress-strain) relation for isotropic linear elastic materials. The most innovative aspect of these two books is Rajen's use of the Elliptical co-ordinate system, which enables a clearer evaluation of the parametric trends of the resulting stress terms.

Students and researchers from a wide range of disciplines from mathematics, physics, and material science to solid mechanics and engineering would find these two books extremely useful for analyzing a wide range of problems, such as glass panel fractures (in car windscreens, for example), tunnelling in soft ground and hard rock, reservoir geomechanics problems associated with carbon dioxide sequestration, etc. In the case of tunnelling and reservoir geomechanics problems, it is not clear how well the finite element and boundary element models capture the deformed state and the risk of localized rupture and subsequent progression of failure; the solutions contained in these two books will be of tremendous help in this regard. Last, but not least, this is a standard problem in steel connections (perforated members bolted to each other) where establishing the strength of the critical crack / tear path is an important design issue. The equations described in these two books could be used to analyze and design plates containing multiple holes via superposition, thereby deriving critical values for the spacing of openings to preclude tear-off prior to global yielding of the steel plate overall.

I commend Rajen for his very detailed and innovative analysis of this age-old problem in classical theory of elasticity using a more general Elliptical co-ordinate system of which the Polar co-ordinate system is a special case. I have no doubt that these two books will be well received and extensively used by the scientific community.

Jitendrapal (Jit) Sharma Ph.D., P.Eng.

Professor and Inaugural Department Chair (2013-2018)

Department of Civil Engineering

York University

Toronto, Ontario, Canada

Contents

List of Tables

List of Figures

Introduction

Stress at the tip of a central elliptical hole in a thin plate (very long in other two dimensions) has not been understood well because of difficulties in determining displacement and deformation gradients when an elliptical shape deforms to form another elliptical shape. It is indeed convenient to use Elliptical co-ordinate system while analysing elliptical holes. However, the grid in an Elliptical co-ordinate system is not fixed; it can shift when an elliptical shape deforms to form another elliptical shape.

In this second book, it is shown that when displacement caused by such a shift is included in the total displacement, it leads us to correct expressions for displacement and deformation gradients and consequently to a much improved analytical solution.

In the first book (titled *Understanding the Elastic Stress Field Around an Elliptical Hole in a Thin Plate (in the deformed configuration)- ISBN 978-1-48359-262-6*), only stress analysis was carried out in the deformed configuration. We did not involve strain-displacement relations, stress-strain relations, and material properties. Hence, the complete solution could not be attained. That is why expressions for stress components presented at the end of Chapter 2 in the first book contained a parameter α_e. How to determine the value of α_e is shown in this second book.

Not to involve strain-displacement relations, stress-strain relations, and material properties in the first book was intentional. By doing so, the dichotomy present in the total solution can be emphasized and the sole impact of geometric non-linearity can be readily interpreted. Furthermore, since strain-displacement relations, stress-strain relations, and material properties are not considered in the first book, the analysis and the results presented in the first book are applicable to all materials.

The first book does not deal with deformation. It shows that, even without dealing with deformation (small or large), the stress field depends on the final configuration. Furthermore, there are special features (e.g. the zero stress point, localized nature of the disturbance, tip stress depends on the shape and not the size of the hole) of such a stress field which are presented in Chapter 3 onward in the first book.

Since the first book does not deal with deformation, it can be a strong reference book for a beginning course on stress analysis where deformation is not introduced (or discussed).

The first book shows how to exploit features of the Elliptical co-ordinate system and the complex analytic functions to obtain expressions for the stress field that contain only one parameter α_e (in comparison with only one parameter in Polar co-ordinate system i.e. radius r). It illustrates how to apply the boundary conditions in the deformed configuration to study the sole impact of geometric non-linearity.

In this second book, we study non-linearity in strain-displacement relations. However, we consider only the linear portion of stress-strain relations. We assume Young's modulus, E, and Poisson's ratio, ν, are constant. Consequently, the analysis and the results presented in this second book are applicable to only linear elastic materials.

Thus, the second book addresses mathematical aspects of only kinematics of this topic and subsequent solutions. Historical perspective and references to any other articles or books are not included. Details of analysis related to stress equilibrium are also not included here. Historical perspective, references, and detailed stress analysis are presented in the first book. However, results derived from the first book which focussed on analysis involving stress equilibrium are used here.

As in the first book, we assume there exists a central elliptical hole in a plate whose length (along X axis) and width (along Y axis) are much larger than its thickness (along Z axis) such that we can treat it as a two dimensional (XY) plane stress problem. It has been shown in the first book (page 32) that at the tip of such an elliptical hole with its major axis along X axis, there exists only vertical stress

$\sigma_{yy} = [S_{yy}] \left[1 + \frac{2a_e}{b_e}\right] - [S_{xx}]$, where S_{xx} and S_{yy} are far field applied stresses (in horizontal i.e. X and vertical i.e. Y directions respectively) while a_e and b_e are semi major and semi minor axes of the elliptical hole at the end (not in the beginning), i.e. in the deformed state. Similarly, at the top of an elliptical hole with its minor axis along Y axis there exists only horizontal stress $\sigma_{xx} = [S_{xx}] \left[1 + \frac{2b_e}{a_e}\right] - [S_{yy}]$. Many researchers do not distinguish between the predeformed (starting) and deformed (ending) values for a and b. Many have formulas for tip stress for uniaxial loading only where terms involving S_{xx} are not involved.

In this second book, first we develop expressions for displacement and deformation gradients when an elliptical shape deforms to form another elliptical shape. Subsequently, these developed expressions for displacement and deformation gradients are linked with the above mentioned expressions for stresses (from the first book) through Hooke's law using four different definitions of strains (Green, engineering, logarithmic, and Almansi) to obtain lengths of semi major axis a_e and semi minor axis b_e of the deformed elliptical hole.

Chapter 1

Useful features of the Elliptical co-ordinate system

1.1 Basic features

In the following discussion, it is assumed that the centers of various shapes under consideration are at the origin. For this study, it is also assumed that there is no applied shear stress i.e. $S_{xy} = S_{yx} = 0$.

In Cartesian co-ordinate system, we need any two of the three 1) a, semi major axis, 2) b, semi minor axis, and 3) c, focal length to define an ellipse since we have one relation,

5

$a^2 - b^2 = c^2$. Normally, in Cartesian co-ordinate system, we use a and b. In Elliptical co-ordinate system, we need any two of the four 1) a, semi major axis, 2) b, semi minor axis, 3) c, focal length, and 4) α, shape, to define an ellipse since we have two relations, $a = c \cosh\alpha$ and $b = c \sinh\alpha$. These two relations are equivalent to one relation, $a^2 - b^2 = c^2$. Normally, in Elliptical co-ordinate system, we use α (to define the shape through $\frac{b}{a}$) and c to define the size of the ellipse.

In Cartesian co-ordinate system, you need x and y to define a point. In Elliptical co-ordinate system, you need α, β, and c to define a point. This is because without knowing c we cannot precisely say where the ellipse representing α lies or where the hyperbola representing β lies. Description of a point defined by α, β, and c in Elliptical system can be converted to Cartesian system through relations $x = c \cosh\alpha \cos\beta$ and $y = c \sinh\alpha \sin\beta$. Hence, $\frac{x}{c \cosh\alpha} = \cos\beta$ and $\frac{y}{c \sinh\alpha} = \sin\beta$ such that $(\frac{x}{c \cosh\alpha})^2 + (\frac{y}{c \sinh\alpha})^2 = \cos^2\beta + \sin^2\beta = 1$. This is equivalent to the equation of an ellipse in Cartesian system $(\frac{x}{a})^2 + (\frac{y}{b})^2 = 1$. Thus, $a = c \cosh\alpha$ and $b = c \sinh\alpha$ provide relationships between two systems for semi major axis and semi minor axis of an ellipse. Hence, α = constant is an ellipse.

When $\alpha = 0$, we obtain $a = c$ and $b = 0$. This is an ellipse with no height, i.e. a straight line. Also, when $\alpha = 0$, $x = c \cosh\alpha \cos\beta = c \cos\beta$ and $y = c \sinh\alpha \sin\beta = 0$. Since β ranges from 0 to π to 2π, x ranges from c to $-c$ to c. Thus, the flat ellipse is a horizontal line of length $2c = 2a$. In addition, when $\beta = 0$, $x = c = a$ and when $\beta = \frac{\pi}{2}$, $x = 0$.

Thus, β curves (hyperbolas as explained below) in the first quadrant are spread over this distance between $x = a$ and $x = 0$. When α is very large, $\sinh\alpha \approx \cosh\alpha$ such that the Elliptical system approximates the Polar system.

Relations $x = c \cosh\alpha \cos\beta$ and $y = c \sinh\alpha \sin\beta$ can also be written as $\frac{x}{c \cos\beta} = \cosh\alpha$ and $\frac{y}{c \sin\beta} = \sinh\alpha$ such that $(\frac{x}{c \cos\beta})^2 - (\frac{y}{c \sin\beta})^2 = \cosh^2\alpha - \sinh^2\alpha = 1$. This is equivalent to the equation of a hyperbola in Cartesian system $(\frac{x}{p})^2 - (\frac{y}{q})^2 = 1$. Hence, $\beta = $ constant is a hyperbola.

In Cartesian co-ordinate system with x and y co-ordinates, $x = $ constant is a vertical line and $y = $ constant is a horizontal line. Similarly, in Elliptical co-ordinate system with α and β co-ordinates, $\alpha = $ constant is an ellipse and $\beta = $ constant is a hyperbola. However, $x = $ constant is a specific line whereas $\alpha = $ constant is not a specific ellipse. This feature is illustrated in Table 1.1.1 and in Figure 1.1.1 where $\alpha = 0.1$ for three ellipses with $c = 0.1$, $c = 0.15$, and $c = 0.2$ respectively. Thus, in addition to α, the focal length, c, is also needed to define a specific ellipse.

Consider three points A $(x,y) = (0.098974, 0.001739)$, B $(x,y) = (0.148460, 0.002609)$, and C $(x,y) = (0.197947, 0.003479)$, shown in Figure 1.1.1. They have the same (α, β) co-ordinates $(0.1, 10^o)$, although they do not lie on the same ellipse. They do not have the same c value. Thus, there is no one to one correspondence between (α, β) co-ordinates and (x,y) co-ordinates.

β	β	α = 0.1						$\frac{x_3}{x_1}$
		$c_1 = 0.1$		$c_2 = 0.15$		$c_3 = 0.2$		
°s	rads	x_1	y_1	x_2	y_2	x_3	y_3	
0	0.000000	0.100500	0.000000	0.150751	0.000000	0.201001	0.000000	2
10	0.174533	0.098974	0.001739	0.148460	0.002609	0.197947	0.003479	2
20	0.349066	0.094440	0.003426	0.141659	0.005139	0.188879	0.006852	2
30	0.523599	0.087036	0.005008	0.130554	0.007513	0.174072	0.010017	2
40	0.698132	0.076988	0.006439	0.115482	0.009658	0.153976	0.012877	2
50	0.872665	0.064600	0.007673	0.096901	0.011510	0.129201	0.015346	2
60	1.047198	0.050250	0.008675	0.075375	0.013012	0.100500	0.017349	2
70	1.221730	0.034373	0.009413	0.051560	0.014119	0.068746	0.018825	2
80	1.396263	0.017452	0.009864	0.026178	0.014797	0.034903	0.019729	2
90	1.570796	0.000000	0.010017	0.000000	0.015025	0.000000	0.020033	

Table 1.1.1: x and y values for thirty points where β varies from 0 to 90 degrees and c equals 0.1, 0.15, and 0.2 while α is 0.1.

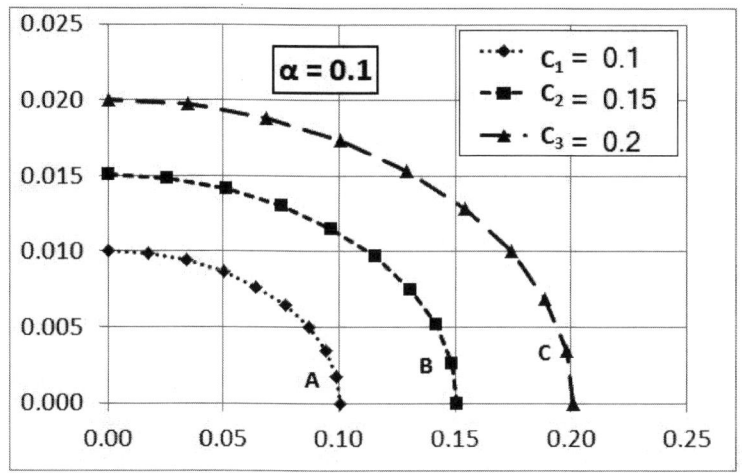

Figure 1.1.1: Three elipses in the first quadrant for $c = 0.1$, 0.15, and 0.2 respectively where $\alpha = 0.1$.

When $\beta = 0$, we obtain $p = c$ and $q = 0$. This is a hyperbola with no height, i.e. a straight line. Also, when $\beta = 0$, $x = c\cosh\alpha\cos\beta = c\cosh\alpha = a$ and $y = c\sinh\alpha\sin\beta = 0$. Since α ranges from 0 to ∞, x ranges from $c = a$ to ∞. Thus, this flat hyperbola is a horizontal line of infinite length, starting from $x = c\cosh\alpha = a$ to $x = \infty$. When $\beta = \pi/2$, $x = 0$ and $y = c\sinh\alpha$. Since α ranges from 0 to ∞, y ranges from 0 to ∞. Hence, this flat hyperbola is a vertical line of infinite length from $y = 0$ to $y = \infty$. Thus, β curves open up from a horizontal line when $\beta = 0$ to a vertical line when $\beta = \pi/2$. Also, vertices of these hyperbolas lie on the horizontal line between $x = c\cosh\alpha = a$ and $x = 0$.

1.2   Effects of change in c

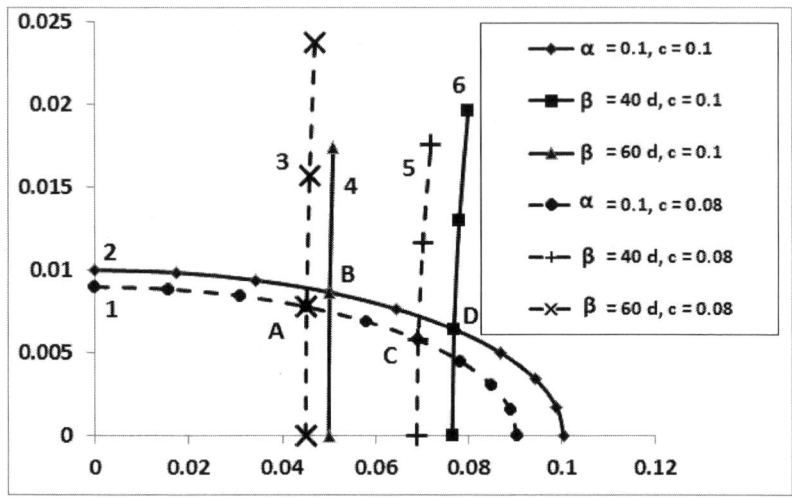

Figure 1.2.1: Ellipses and hyperbolas come closer to the center when c is decreased from 0.1 to 0.08.

As shown in Figure 1.1.1, as c decreases, ellipses come closer to the center although there is no change in α or β. Similarly, hyperbolas come closer to the center. Figure 1.2.1 displays this effect.

In Figure 1.2.1, point B is at the intersection of $\alpha = 0.1$ (graph 2) and $\beta = 60°$ (graph 4). Similarly, point D is at the intersection of $\alpha = 0.1$ (graph 2) and $\beta = 40°$ (graph 6). Both points lie on the ellipse (graph 2) for which $\alpha = 0.1$ and $c = 0.1$. When c decreases from 0.1 to 0.08, all three graphs (graph 2, graph 4, and graph 6) move closer to the center. These three new graphs (graph 1, graph 3, and graph 5) are shown using dashed lines. Accordingly, point B moves to point A and point D moves to point C. Points A and B have the same co-ordinates $(0.1, 60°)$. Similarly, points C and D have the same co-ordinates $(0.1, 40°)$. However, points A, B, C, and D have different (x, y) co-ordinates.

Thus, when c decreases, α curves (ellipses) and β curves (hyperbolas) move towards the center. Conversely, when c increases, α curves (ellipses) and β curves (hyperbolas) move away from the center. We conclude that in Cartesian system the x-y grid is fixed, but in Elliptical system the α-β grid is not fixed; it is linked to c. When $c = 0$, focal points and center merge, ellipses become circles, and vertices of hyperbolas merge into the origin. Hyperbolas become straight lines (radii) converging at the center. Elliptical co-ordinate system becomes Polar co-ordinate system.

Larger c value implies $\alpha = 0.0$ is a longer line (a flat ellipse of length $2c$) and ellipses $\alpha > 0.0$ are pushed away from the center. Larger c value also implies the value of $a = c \cosh\alpha$ is larger and vertices of β curves are spread out over this longer distance. Furthermore, the distances between β curves are smallest at $\alpha = 0.0$. As α increases, the distances between β curves (along $\alpha = $ constant ellipses) increase. The shortest distance between $\beta = 0$ (a horizontal line starting at $x = c \cosh\alpha = a$ and moving away from the center) and $\beta = \frac{\pi}{2}$ (a vertical line starting at $x = 0$ and moving up from the center) is $c = a$ (from the origin to $x = c = a$, $y = 0$) when $\alpha = 0.0$.

					$c = 0.1$				
		$\alpha_1 = 0.1$		$\alpha_2 = 0.15$		$\alpha_3 = 0.2$			
β	β	x_1	y_1	x_2	y_2	x_3	y_3		$\frac{x_3}{x_1}$
°s	rads								
0	0.000000	0.100500	0.000000	0.101127	0.000000	0.102007	0.000000		1.015
10	0.174533	0.098974	0.001739	0.099591	0.002615	0.100457	0.003496		1.015
20	0.349066	0.094440	0.003426	0.095028	0.005150	0.095855	0.006886		1.015
30	0.523599	0.087036	0.005008	0.087579	0.007528	0.088340	0.010067		1.015
40	0.698132	0.076988	0.006439	0.077468	0.009678	0.078142	0.012942		1.015
50	0.872665	0.064600	0.007673	0.065003	0.011534	0.065569	0.015423		1.015
60	1.047198	0.050250	0.008675	0.050564	0.013039	0.051003	0.017436		1.015
70	1.221730	0.034373	0.009413	0.034588	0.014148	0.034888	0.018919		1.015
80	1.396263	0.017452	0.009864	0.017561	0.014828	0.017713	0.019828		1.015
90	1.570796	0.000000	0.010017	0.000000	0.015056	0.000000	0.020134		

Table 1.2.1: x and y values for thirty points where β varies from 0 to 90 degrees and α equals 0.1, 0.15, and 0.2 while c is 0.1.

1.3   Effects of change in α

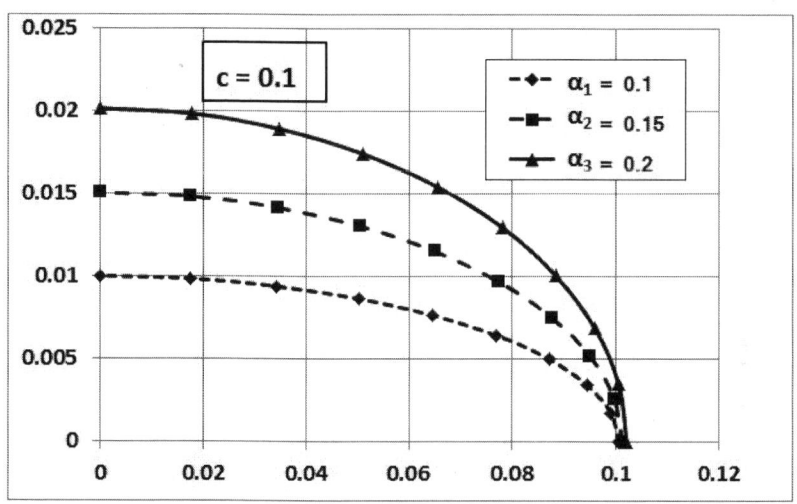

Figure 1.3.1: Three elipses in the first quadrant for $\alpha = 0.1$, 0.15, and 0.2 respectively where $c = 0.1$.

Whenever c changes; since $a^2 - b^2 = c^2$, a changes, or b changes, or both a and b change. If such a change maintains the ratio $\frac{b}{a}$ same as before then α will not change. Otherwise, α will also change. For example, if a reduces from 6 to 3 while b reduces from 2 to 1 then $\frac{b}{a}$ remains at $\frac{1}{3}$ whereas c reduces from $4\sqrt{2}$ to $2\sqrt{2}$. Since $\frac{b}{a}$ remains at $\frac{1}{3}$, α does not change.

However, in this chapter, we study effects of changing only c (as in Figure 1.2.1) or only α. Hence, we now examine the effect of changes in only α. In Table 1.1.1, we selected

13

three values for c, keeping α constant. In Table 1.2.1, we select three values for α, keeping c constant. For both tables, we selected ten β values to study the effects at various points on the ellipses. Thus, thirty points are plotted in Figure 1.3.1 to obtain three graphs. Since c does not change, the α-β grid is not affected.

We note that when $\beta = 0$, $\sin\beta = 0$, hence $y = 0$. When $\beta = \pi/2$, $\cos\beta = 0$, hence $x = 0$. Consequently, for each graph in Figure 1.3.1, as β increases from 0 to $\pi/2$, x decreases from $x = c \cosh\alpha$ to 0 whereas y increases from 0 to $y = c \sinh\alpha$. An increase in α causes both $x = c \cosh\alpha \cos\beta$ and $y = c \sinh\alpha \sin\beta$ to increase. Therefore, graph for $\alpha = 0.2$ is above the graph for $\alpha = 0.15$.

However, at low values of α, $\cosh\alpha$ is much higher than $\sinh\alpha$. Hence, $x = c \cosh\alpha$ values at $\beta = 0^\circ$ are much higher than $y = c \sinh\alpha$ values at $\beta = 90^\circ$. For example, x_1 at $\beta = 0^\circ$ is 0.1005 whereas y_1 at $\beta = 90^\circ$ is 0.010017.

Please note that, in Figure 1.3.1, the scale for x-axis is not the same as the scale for y-axis.

Chapter 2

Displacements in the Elliptical co-ordinate system

2.1 Effects of changes in c and α

Consider the point $(0.1, \pi/3)$ in Elliptical co-ordinate system. When $c = 0.1$,

$$x = c \cosh\alpha \, \cos\beta = 0.1 \cosh(0.1) \cos(\pi/3)$$
$$= (0.1)[1.005004][0.5] = 0.050250 \text{ and}$$
$$y = c \sinh\alpha \, \sin\beta = 0.1 \sinh(0.1) \sin(\pi/3)$$
$$= (0.1)[0.100167][0.866025] = 0.008675.$$

However, when $c = 0.09$,

$$x = c \cosh\alpha \, \cos\beta = 0.09 \cosh(0.1) \, \cos(\pi/3)$$
$$= (0.09)[1.005004][0.5] = 0.045225 \text{ and}$$
$$y = c \sinh\alpha \, \sin\beta = 0.09 \sinh(0.1) \, \sin(\pi/3)$$
$$= (0.09)[0.100167][0.866025] = 0.007807.$$

Thus, when $c = 0.1$, the point $(0.1, \pi/3)$ in Elliptical co-ordinate system represents the point $(0.050250, 0.008675)$ in Cartesian co-ordinate system, and when $c = 0.09$, the same point $(0.1, \pi/3)$ in Elliptical co-ordinate system represents the point $(0.045225, 0.007807)$ in Cartesian co-ordinate system.

This is somewhat identical to a passenger sitting behind the driver in a car. His geographical co-ordinates change when the car moves (equivalent to a change in c) to a different location although the passenger is still at the same seat (same α and β co-ordinates) in the car, behind the driver.

This is also identical to a little bug sitting on a spike of a partially open umbrella. When the umbrella is opened fully, co-ordinates of the bug change although it may be still sitting at the same place on that spike.

Here, although the centers of the shape and α and β co-ordinates in the elliptical framework do not change, the displacement is strictly because of shifting focal points which cause a shift in the grid.

Now, suppose α increases from $\alpha = 0.1$ to $\alpha = 0.12$, when $c = 0.09$, then

$$x = c \cosh\alpha \, \cos\beta = 0.09 \cosh(0.12) \, \cos(\pi/3)$$

$$= (0.09)[1.007209][0.5] = 0.045324 \text{ and}$$

$$y = c \sinh\alpha \, \sin\beta = 0.09 \sinh(0.12) \, \sin(\pi/3)$$

$$= (0.09)[0.110222][0.866025] = 0.009376.$$

This is somewhat identical to the passenger moving from behind the driver seat to the next seat when the car is not moving or the bug on the spike making a move when the umbrella is still.

Here, although the center of the shape (car or umbrella which are still) and the focal points as well as β co-ordinate in the elliptical framework do not change, the displacement is strictly because of change in α value.

An increase in α value causes an increase in x from 0.045225 to 0.045324 as well as an increase in y from 0.007807 to 0.009376.

2.2 Components of a displacement vector

Consider a horizontal slit of starting focal length of 0.1. Hence, $c_s = 0.1$ and $\alpha_s = 0.0$. Suppose it deforms under the influence of far field stresses to form an ellipse such that c_e = 0.09 and $\alpha_e = 0.11$. The material point at the right tip (where $\beta = 0$) of the undeformed slit is at

$$x_s = c_s \cosh\alpha_s \, \cos\beta = 0.1 \cosh(0.0) \, \cos(0)$$

$$= (0.1)[1][1] = 0.1 \text{ and}$$

$y_s = c_s \sinh\alpha_s \sin\beta = 0.1 \sinh(0.0) \sin(0)$

$\qquad = (0.1)[0.0][0.0] = 0.0.$

This tip after deformation moves to

$x_e = c_e \cosh\alpha_e \cos\beta = 0.09 \cosh(0.11) \cos(0)$

$\qquad = 0.090545$ and

$y_e = c_e \sinh\alpha_e \sin\beta = 0.09 \sinh(0.11) \sin(0)$

$\qquad = (0.09)[\sinh(0.11)][0.0] = 0.0.$

Here, subscript s refers to the starting (predeformed) position and subscript e refers to the ending (deformed) position.

Thus, the right tip moves from (0.1,0) to (0.090545,0). The slope of this displacement vector is zero. The magnitude is 0.1 - 0.090545 = 0.009455. Thus, the displacement vector is normal to the deformed surface of the hole. It is pointing west (towards the center).

Now, let us breakdown this displacement vector in two components:

1) due to a change in c from 0.1 to 0.09. Here,

$x_1 = c_e \cosh\alpha_s \cos\beta = 0.09 \cosh(0.0) \cos(0)$

$\qquad = 0.09$ and

$y_1 = c_e \sinh\alpha_s \sin\beta = 0.09 \sinh(0.0) \sin(0)$

$\qquad = (0.09)[0.0][0.0] = 0.0.$

Thus, this component of displacement is from (0.1,0) to (0.09,0). The slope of this component of the displacement is zero. The magnitude is 0.1 - 0.09 = 0.01. This component of the displacement is along major axis and is pointing west (towards the center).

2) due to a change in α from 0.0 to 0.11.

This component of displacement is from (0.09,0) to (0.090545,0.0). The slope of this component of the displacement is zero. The magnitude is 0.000545 This component of displacement is strictly due to a change in α. This component of the displacement is along major axis and is pointing east (away from the center). As a result, the total displacement is 0.01 - 0.000545 = 0.009455, pointing west (towards the center).

At the tip, component due to change in c as well as component due to a change in α, both are horizontal.

We note that for the tip, $\beta = 0$, hence both y_s and y_e are zero. Therefore, the displacement at the tip has no vertical component . This is also obvious from symmetry point of view.

Now, consider the material point at $\beta = \pi/180$ on the same ellipse. For this point,

$$x_s = c_s \cosh\alpha_s \cos\beta = 0.1 \cosh(0.0) \cos(\pi/180)$$
$$= (0.1)[1][0.999848] = 0.0999848 \text{ and}$$
$$y_s = c_s \sinh\alpha_s \sin\beta = 0.1 \sinh(0.0) \sin(\pi/180)$$
$$= 0.0.$$

This material point after deformation moves to

$$x_e = c_e \cosh\alpha_e \cos\beta = 0.09 \cosh(0.11) \cos(\pi/180)$$

$$= 0.09[1.006056][0.999848] = 0.090531 \text{ and}$$

$$y_e = c_e \sinh\alpha_e \sin\beta = 0.09 \sinh(0.11) \sin(\pi/180)$$

$$= (0.09)[0.1102][0.017453] = 0.0001731.$$

This material point moves from (0.0999848,0) to (0.090531,0.0001731).

Thus, the material point just left of the right tip moves mostly horizontally left and vertically slightly up. The slope of the displacement is negative and equals (0.0001731-0)/(0.090531-0.0999848) = (0.0001731)/(-0.0094538) = - 0.01831. The magnitude is $\sqrt{((0.0001731)^2 + (-0.0094538)^2)} = 0.009454328$. The displacement is not normal to the surface of the hole. It is in mostly west slightly north direction.

Now, let us breakdown this displacement in two components :

1) due to a change in c from 0.1 to 0.09. Here,

$$x_1 = c_e \cosh\alpha_s \cos\beta = 0.09 \cosh(0.0) \cos(\pi/180)$$

$$= 0.09[1][0.999848] = 0.0899863 \text{ and}$$

$$y_1 = c_e \sinh\alpha_s \sin\beta = 0.09 \sinh(0.0) \sin(\pi/180)$$

$$= (0.09)[0.0][0.017452] = 0.0.$$

This component of displacement is from (0.0999848,0) to (0.0899863,0). The slope of this component of the displacement is zero. The magnitude is 0.0999848-0.0899863 = 0.0099985 This component of the displacement is along major axis and is pointing west.

2) due to a change in α from 0.0 to 0.11. This component of displacement is from (0.0899863,0) to (0.090531,0.0001731).

The slope of this component of the displacement is positive and equals $(0.0001731 - 0)/(0.090531 - 0.0899863) = (0.0001731)/(0.0005448) = 0.31771$. The magnitude is $\sqrt{((0.0001731)^2 + (0.0005448)^2)} = 0.000572$. This component of displacement is strictly due to a change in α. It is pointing away from the center (northeast) due to increase in α. However, it is very small in magnitude as compared to the component due to change in c.

2.3 Analysis of displacements of points along the surface of an elliptical hole

Similar calculations were carried out for various β values for many cases. In particular, five representative cases which are listed below were studied in detail.

Case 1 - A horizontal slit experiencing contraction along major axis and expansion along minor axis,

Case 2 - A horizontal elliptical hole experiencing contraction along major axis and expansion along minor axis,

Case 3 - A horizontal elliptical hole experiencing expansion along major axis and expansion along minor axis,

Case 4 - A horizontal elliptical hole experiencing expansion along major axis and contraction along minor axis, and

Case 5 - A horizontal elliptical hole experiencing contraction along major axis and contraction along minor axis.

		Starting		Ending		Change c		Change α	
		$c_s = 0.1$	$\alpha_s = 0.05$	$c_e = 0.09$	$\alpha_e = 0.11$	$c_e = 0.09$	$\alpha_s = 0.05$	$c_s = 0.1$	$\alpha_e = 0.11$
β °s	β rads	x_s	y_s	x_e	y_e	x_2	y_2	x_3	y_3
0	0.0000	0.1001	0.0000	0.0905	0.0000	0.0901	0.0000	0.1006	0.0000
10	0.1745	0.0986	0.0009	0.0892	0.0017	0.0887	0.0008	0.0991	0.0019
20	0.3491	0.0941	0.0017	0.0851	0.0034	0.0847	0.0015	0.0945	0.0038
30	0.5236	0.0867	0.0025	0.0784	0.0050	0.0780	0.0023	0.0871	0.0055
40	0.6981	0.0767	0.0032	0.0694	0.0064	0.0690	0.0029	0.0771	0.0071
50	0.8727	0.0644	0.0038	0.0582	0.0076	0.0579	0.0034	0.0647	0.0084
60	1.0472	0.0501	0.0043	0.0453	0.0086	0.0451	0.0039	0.0503	0.0095
70	1.2217	0.0342	0.0047	0.0310	0.0093	0.0308	0.0042	0.0344	0.0104
80	1.3963	0.0174	0.0049	0.0157	0.0098	0.0156	0.0044	0.0175	0.0109
90	1.5708	0.0000	0.0050	0.0000	0.0099	0.0000	0.0045	0.0000	0.0110

Table 2.3.1: x and y values for forty points where β varies from 0 to 90 degrees and α equals 0.05 or 0.11 while c equals 0.1 or 0.09.

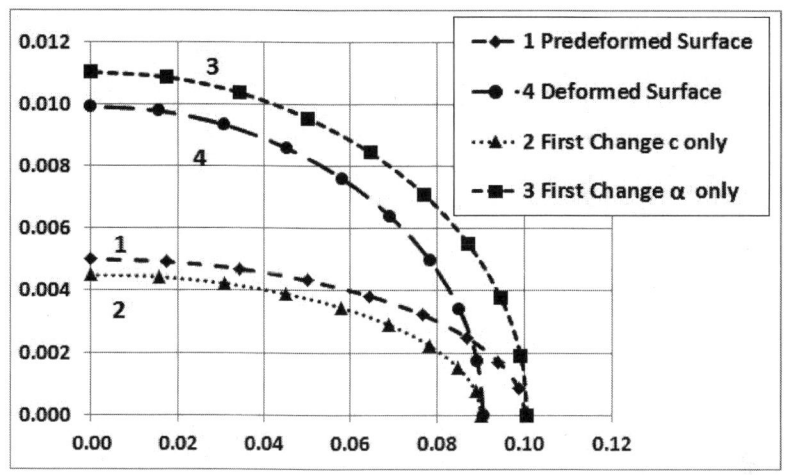

Figure 2.3.1: Graphs showing different contours.

However, for the sake of brevity, details for only Case 2 are presented below. Other cases are discussed in Chapter 11.

There are five vertical sections in Table 2.3.1. The first section shows selected ten β values in degrees and radians. The second section shows x_s and y_s values corresponding to selected ten β values for the predeformed (starting) position of the contour of the hole described by $c_s = 0.1$ and $\alpha_s = 0.05$. These x_s and y_s values are used to plot Graph 1 in Figure 2.3.1. The third section shows x_e and y_e values corresponding to selected ten β values for the deformed (end) position of the contour of the hole described by $c_e = 0.09$ and $\alpha_e = 0.11$. These x_e and y_e values are used to plot Graph 4 in Figure 2.3.1. The fourth section shows x_2 and y_2 values corresponding to selected ten β values for the intermediate position (if only c changed first) of the contour of the hole described by $c_e = 0.09$ and $\alpha_s = 0.05$. These x_2 and y_2 values are used to plot Graph 2 in Figure 2.3.1. The fifth section shows x_3 and y_3 values corresponding to selected ten β values for the intermediate position (if only α changed first) of the contour of the hole described by $c_s = 0.1$ and $\alpha_e = 0.11$. These x_3 and y_3 values are used to plot Graph 3 in Figure 2.3.1.

In Figure 2.3.1, we observe that a reduction in c brings the contour closer to the center. Whereas, an increase in α pushes the contour away from the center. Furthermore, if only c is reduced first from $c_s = 0.1$ to $c_e = 0.09$, leaving $\alpha_s = 0.05$, the predeformed contour shown by Graph 1 moves to the contour shown by Graph 2. Subsequently, if $\alpha_s = 0.05$ is increased to $\alpha_e = 0.11$, leaving $c_e = 0.09$, we reach the deformed contour shown by Graph 4. On the other hand, if only α is increased first from $\alpha_s = 0.05$ to $\alpha_e = 0.11$, leaving $c_s = 0.1$, the predeformed contour shown by Graph 1 moves to the contour shown by Graph 3. Subsequently, if c is reduced from $c_s = 0.1$ to $c_e = 0.09$, leaving $\alpha_e = 0.11$, we reach the deformed contour shown by Graph 4. Thus, by changing only c or α at a time, we are able to study its effect at various points on the contour, as shown below.

Horizontal and vertical components of displacements

β	β	$c_s = 0.1$ $\alpha_s = 0.05$		$c_e = 0.09$ $\alpha_e = 0.11$	
$^\circ s$	rads	$x_e - x_s$	$y_e - y_s$	Magnitude	Slope
0	0.0000	-0.0096	0.0000	0.0096	0.0000
10	0.1745	-0.0094	0.0009	0.0095	-0.0905
20	0.3491	-0.0090	0.0017	0.0092	-0.1868
30	0.5236	-0.0083	0.0025	0.0087	-0.2964
40	0.6981	-0.0073	0.0032	0.0080	-0.4308
50	0.8727	-0.0062	0.0038	0.0072	-0.6118
60	1.0472	-0.0048	0.0043	0.0064	-0.8892
70	1.2217	-0.0033	0.0046	0.0057	-1.4104
80	1.3963	-0.0017	0.0048	0.0051	-2.9114
90	1.5708	0.0000	0.0049	0.0049	∞

Table 2.3.2: Values for horizontal and vertical components of displacements as well as their magnitudes and slopes at ten points where β varies from 0 to 90 degrees.

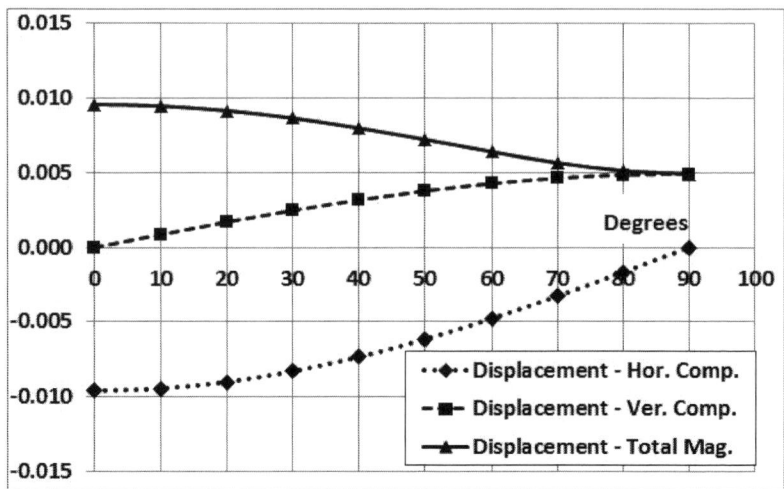

Figure 2.3.2: The top graph shows the magnitudes of displacements at ten points where β varies from 0 to 90 degrees. The middle and bottom graphs show vertical and horizontal components of displacements at these ten points.

Before we study the effect of changing c or α, we breakdown the ten displacement vectors into their horizontal and vertical components , as shown in Table 2.3.2 and Figure 2.3.2. From this table and the figure, we can readily conclude that at the tip where $\beta = 0^o$, there is no vertical component since $y = c \sinh\alpha \, \sin\beta = 0$. Similarly, at the top where $\beta = 90^o$, there is no horizontal component since $x = c \cosh\alpha \, \cos\beta = 0$. This conclusion is also compatible from the symmetry point of view.

26

We observe that, at all points, the horizontal component is negative whereas the vertical component is positive. As a result, the slope of the displacement vector, at all points, is negative. Furthermore, the slope is zero at the tip since the vertical component is zero. However, the slope increases in magnitude yet remains negative as we move from the tip to the top. Consequently, there is significant movement towards the center near the tip. On the other hand, there is significant movement away from the center near the top.

Furthermore, from the Figure 2.3.2 as well as the Table 2.3.2, we observe that the magnitude of the displacement declines as we move from the tip to the top. The displacement vector is always perpendicular to the surface of the hole at the tip and the top. However, the displacement vector may not be perpendicular to the surface at any other point along the surface of the hole.

Now, we breakdown the total displacement at each of the ten points into components due to change in c first and then in α. There are five vertical sections in Table 2.3.3. The first section shows selected ten β values in degrees and radians. The second section shows the horizontal and the vertical components of displacements due to change in c first i.e. $x_2 - x_s$ and $y_2 - y_s$ values corresponding to selected ten points on the predeformed (starting) position of the contour of the hole (Graph 1 in Figure 2.3.1) described by $c_s = 0.1$ and $\alpha_s = 0.05$ to the contour (Graph 2 in Figure 2.3.1) described by $c_e = 0.09$ and $\alpha_s = 0.05$. The third section shows the magnitudes and slopes of these ten displacements. These magnitudes are plotted with dotted line in Figure 2.3.3.

β	β	Starting $c_s = 0.1$, $\alpha_s = 0.05$		Change c first $c_e = 0.09$, $\alpha_s = 0.05$		Change α later $c_e = 0.09$, $\alpha_e = 0.11$		Ending $c_e = 0.09$, $\alpha_e = 0.11$	
$°_s$	rads	$x_2 - x_s$	$y_2 - y_s$	Mag.	Slope	$x_e - x_2$	$y_e - y_2$	Mag.	Slope
0	0.0000	-0.0100	0.0000	0.0100	0.0000	0.0004	0.0000	0.0004	0.0000
10	0.1745	-0.0099	-0.0001	0.0099	0.0088	0.0004	0.0009	0.0010	2.2088
20	0.3491	-0.0094	-0.0002	0.0094	0.0182	0.0004	0.0019	0.0019	4.5593
30	0.5236	-0.0087	-0.0003	0.0087	0.0288	0.0004	0.0027	0.0027	7.2323
40	0.6981	-0.0077	-0.0003	0.0077	0.0419	0.0003	0.0035	0.0035	10.511
50	0.8727	-0.0064	-0.0004	0.0064	0.0595	0.0003	0.0042	0.0042	14.929
60	1.0472	-0.0050	-0.0004	0.0050	0.0865	0.0002	0.0047	0.0047	21.697
70	1.2217	-0.0034	-0.0005	0.0035	0.1373	0.0001	0.0051	0.0051	34.417
80	1.3963	-0.0017	-0.0005	0.0018	0.2833	0.0001	0.0053	0.0053	71.042
90	1.5708	0.0000	-0.0005	0.0005	∞	0.0000	0.0054	0.0054	∞

Table 2.3.3: Horizontal and vertical components of displacements due to changes first in c and then in α as well as their magnitudes and slopes for ten values of β ranging from 0 to 90.

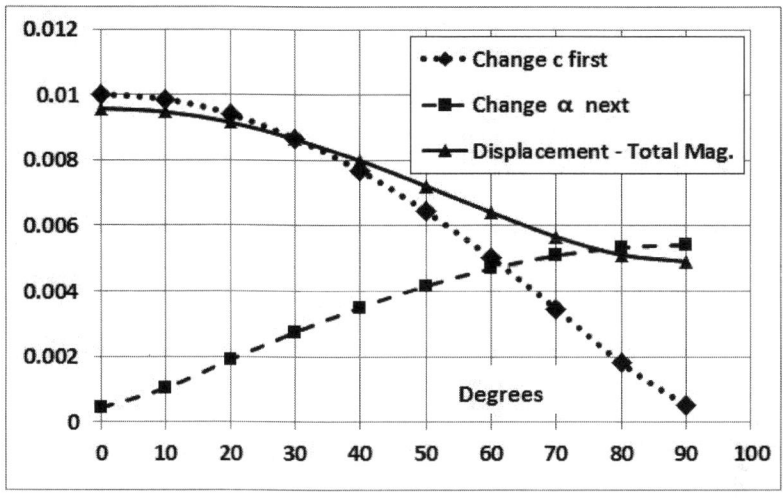

Figure 2.3.3: Three graphs showing the total displacements and their components due to change in c first and subsequent change in α.

The fourth section shows the horizontal and the vertical components of displacements due to subsequent change in α i.e. $x_e - x_2$ and $y_e - y_2$ values corresponding to selected ten points on the contour (Graph 2 in Figure 2.3.1) described by $c_e = 0.09$ and $\alpha_s = 0.05$ to the deformed (end) position of the contour (Graph 4 in Figure 2.3.1) of the hole described by $c_e = 0.09$ and $\alpha_e = 0.11$. The fifth section shows the magnitudes and slopes of these ten displacements. These magnitudes are plotted with dashed line in Figure 2.3.3 where the graph with solid line represents the total displacements.

β	β	Starting $c_s = 0.1$ $\alpha_s = 0.05$		Change α first $c_s = 0.1$ $\alpha_e = 0.11$		Change c later $c_e = 0.09$ $\alpha_e = 0.11$		Ending $c_e = 0.09$ $\alpha_e = 0.11$	
°s	rads	$x_3 - x_s$	$y_3 - y_s$	Mag.	Slope	$x_e - x_3$	$y_e - y_3$	Mag.	Slope
0	0.0000	0.0005	0.0000	0.0005	0.0000	-0.0101	0.0000	0.0101	0.0000
10	0.1745	0.0005	0.0010	0.0011	2.2088	-0.0099	-0.0002	0.0099	0.0193
20	0.3491	0.0005	0.0021	0.0021	4.5593	-0.0095	-0.0004	0.0095	0.0399
30	0.5236	0.0004	0.0030	0.0030	7.2323	-0.0087	-0.0006	0.0087	0.0063
40	0.6981	0.0004	0.0039	0.0039	10.511	-0.0077	-0.0007	0.0077	0.0919
50	0.8727	0.0003	0.0046	0.0046	14.929	-0.0065	-0.0008	0.0065	0.1306
60	1.0472	0.0002	0.0052	0.0052	21.697	-0.0050	-0.0010	0.0051	0.1898
70	1.2217	0.0002	0.0057	0.0057	34.417	-0.0034	-0.0020	0.0036	0.3010
80	1.3963	0.0001	0.0059	0.0059	71.042	-0.0017	-0.0011	0.0021	0.6213
90	1.5708	0.0000	0.0060	0.0060	∞	0.0000	-0.0011	0.0011	∞

Table 2.3.4: Horizontal and vertical components of displacements due to changes first in α and then in c as well as their magnitudes and slopes for ten values of β ranging from 0 to 90.

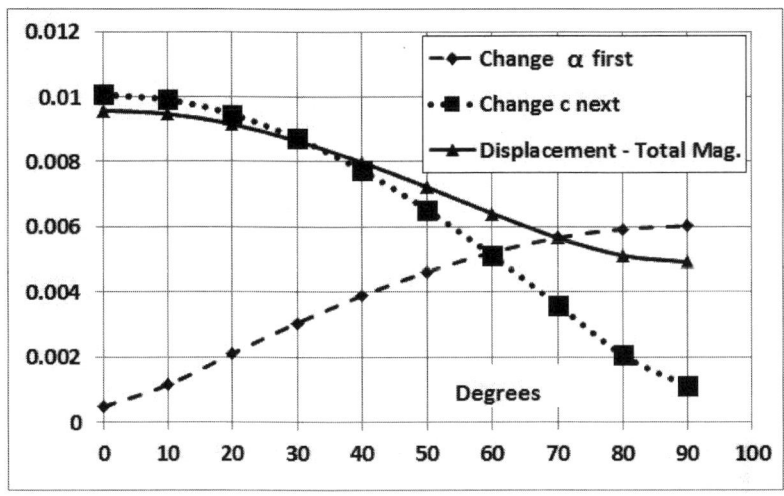

Figure 2.3.4: Three graphs showing the total displacements and their components due to change in α first and subsequent change in c.

Next, we breakdown the total displacement at each of the ten points into components due to change in α first and then in c. Table 2.3.4 is similar to Table 2.3.3 with five vertical sections. The first section shows selected ten β values in degrees and radians. The second section shows the horizontal and the vertical components of displacements due to change in α first i.e. $x_3 - x_s$ and $y_3 - y_s$ values corresponding to selected ten points on the predeformed (starting) position of the contour of the hole (Graph 1 in Figure 2.3.1) described by $c_s = 0.1$ and $\alpha_s = 0.05$ to the contour (Graph 3 in Figure 2.3.1) described by $c_s = 0.1$ and $\alpha_e = 0.11$. The third section shows the magnitudes and slopes of these ten displacements.

These magnitudes are plotted with dashed line in Figure 2.3.4.

The fourth section shows the horizontal and the vertical components of displacements due to subsequent change in c i.e. $x_e - x_3$ and $y_e - y_3$ values corresponding to selected ten points on the contour (Graph 3 in Figure 2.3.1) described by $c_s = 0.1$ and $\alpha_e = 0.11$ to the deformed (end) position of the contour (Graph 4 in Figure 2.3.1) of the hole described by $c_e = 0.09$ and $\alpha_e = 0.11$. The fifth section shows the magnitudes and slopes of these ten displacements. These magnitudes are plotted with dotted line in Figure 2.3.4 where the graph with solid line represents the total displacements.

It can be readily seen from Figure 2.3.3 and Figure 2.3.4 that the component due to c plays a major role in the vicinity of the tip whereas the component due to α plays a major role in the vicinity of the top. A comparison between Table 2.3.3 and Table 2.3.4 as well as between Figure 2.3.3 and Figure 2.3.4 reveals that it does not matter whether the component due to c is considered first or the component due to α is considered first, the magnitudes and the slopes of these components remain the same.

Figure 2.3.5 shows three displacement vectors PR, LN, and AC starting at points $P(\beta = 30^o)$, $L(\beta = 40^o)$, and $A(\beta = 50^o)$ on the predeformed contour and ending at R, N, and C on the deformed contour respectively. We observe that slopes of these vectors increase as β increases while remaining negative. Actual magnitudes and slopes of these vectors are shown in Table 2.3.1.

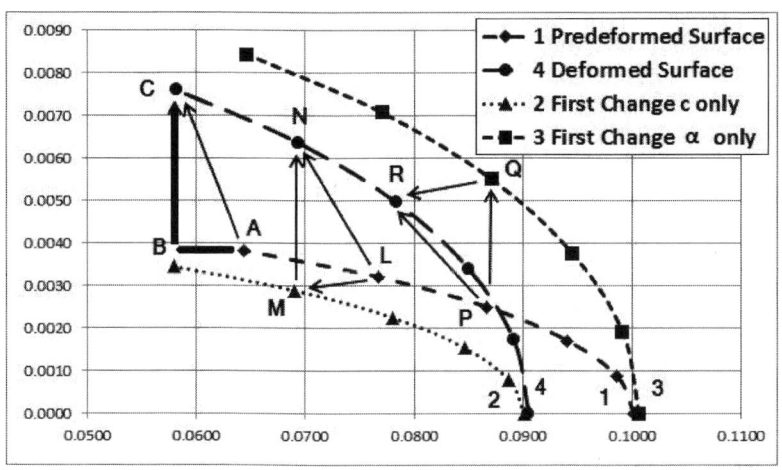

Figure 2.3.5: Three displacement vectors with their components.

Vectors AB and BC are the horizontal and vertical components of the displacement vector AC. Vectors LM and MN are the components due to first change in c and subsequent change in α of the displacement vector LN. Whereas vectors PQ and QR are the components due to first change in α and subsequent change in c of the displacement vector PR. We note that components LM and QR point towards the center due to a decrease in c from $c_s = 0.1$ to $c_e = 0.09$. Whereas components MN and PQ point away from the center due to an increase in α from $\alpha_s = 0.05$ to $\alpha_e = 0.11$.

2.4   Summary

As mentioned in the beginning of Section 2.3, many cases were studied although detailed analysis for the case where α increases and c decreases is presented here.

We note three important features as follows :

1) At the tip, the displacement is always horizontal whereas at the top, the displacement is always vertical. Various studies by some researchers have used this feature on the basis of symmetry although they have not used the Elliptical coordinate system.

2) Along other points on the contour, the displacement may not be perpendicular to the contour. Some researchers have misunderstood this feature.

3) Displacements have two components, due to a change in α and due to a change in c. Many researchers have not recognised the component due to a change in c. Consequently, they have been led to erroneous results.

Chapter 3

Co-ordinate transformations

Expressions developed below for $\tanh\alpha$ and $\tan\beta$ in terms of x, y, and c appear in Appendix A on page 548 of an article by Weertman [2005][1]. Also, expressions developed below for f_1, $\sinh^2\alpha$, and $\sin^2\beta$ are identical to the expressions presented by Maugis [1992] [2] in Appendix 1 on page 255 of his article. Dr. Amin Eshraghi provided me these two (and some more) references. Please note that these two references are listed at the end of this chapter. These expressions may also be presented in other papers and/or textbooks.

However, I could not find detailed derivations for these expressions as well as descriptions of the nature of f_1, f_2, and f_3 which I have presented in this chapter for completeness, better understanding of Elliptical co-ordinate system, and ready reference.

In section 1.2, we noted that when $c = 0$, Elliptical co-ordinate system becomes Polar co-ordinate system. Hence, we consider Polar co-ordinate system first.

3.1 r in terms of x and y

In Polar co-ordinate system,

$$x = r \cos\theta \tag{3.1.1}$$

and

$$y = r \sin\theta. \tag{3.1.2}$$

Hence, $x^2 + y^2 = r^2 \cos^2\theta + r^2 \sin^2\theta = r^2$. And,

$$r = \sqrt{x^2 + y^2} \tag{3.1.3}$$

We note that since r cannot be negative, we accept only positive values for r.

3.2 θ in terms of x and y

From equations (3.1.1) and (3.1.2), we obtain

$\cos\theta = \frac{x}{r}$ and $\sin\theta = \frac{y}{r}$ where $r > 0$.

Hence,

$$\tan\theta = \frac{\sin\theta}{\cos\theta} = \frac{\frac{y}{r}}{\frac{x}{r}} = \frac{y}{x}. \tag{3.2.1}$$

3.3 c in terms of x, y, and α

In Elliptical co-ordinate system,

$$x = c \cosh\alpha \cos\beta \tag{3.3.1}$$

and

$$y = c \sinh\alpha \sin\beta. \tag{3.3.2}$$

From equations (3.3.1) and (3.3.2), we obtain

$\cos^2\beta = \frac{x^2}{c^2 \cosh^2\alpha}$ and $\sin^2\beta = \frac{y^2}{c^2 \sinh^2\alpha}$.

Addition of these two expressions yields

$\cos^2\beta + \sin^2\beta = 1 = \frac{x^2}{c^2 \cosh^2\alpha} + \frac{y^2}{c^2 \sinh^2\alpha}$.

Hence, $c^2 = \frac{x^2}{\cosh^2\alpha} + \frac{y^2}{\sinh^2\alpha}$.

And,

$$c = \sqrt{\frac{x^2}{\cosh^2\alpha} + \frac{y^2}{\sinh^2\alpha}} \tag{3.3.3}$$

We note that since c cannot be negative, we accept only positive values for c.

3.4 c in terms of x, y, and β

From equations (3.3.1) and (3.3.2), we obtain

$\cosh^2\alpha = \frac{x^2}{c^2\cos^2\beta}$ and $\sinh^2\alpha = \frac{y^2}{c^2\sin^2\beta}$.

Since $\cosh^2\alpha - \sinh^2\alpha = 1$, we obtain $\frac{x^2}{c^2\cos^2\beta} - \frac{y^2}{c^2\sin^2\beta} = 1$.

Hence, $c^2 = \frac{x^2}{\cos^2\beta} - \frac{y^2}{\sin^2\beta}$.

And,

$$c = \sqrt{\frac{x^2}{\cos^2\beta} - \frac{y^2}{\sin^2\beta}} \qquad (3.4.1)$$

Once again, since c cannot be negative, we accept only positive values for c.

3.5 α in terms of x, y, and c

Squaring both sides of equation (3.3.1) and equation (3.3.2), we obtain

$x^2 = c^2\cosh^2\alpha\cos^2\beta$ and $y^2 = c^2\sinh^2\alpha\sin^2\beta$.

We can rewrite x^2 as follows :

$x^2 = c^2\cosh^2\alpha\cos^2\beta$

$\quad = c^2\left(1 + \sinh^2\alpha\right)\left(1 - \sin^2\beta\right)$

$\quad = c^2\left(1 + \sinh^2\alpha - \sin^2\beta - \sinh^2\alpha\sin^2\beta\right)$

$\quad = c^2\left(1 + \sinh^2\alpha - \sin^2\beta\right) - c^2\sinh^2\alpha\sin^2\beta$

$$= c^2 \left(1 + \sinh^2\alpha - \sin^2\beta\right) - y^2.$$

Therefore, $x^2 + y^2 = c^2 \left(1 + \sinh^2\alpha - \sin^2\beta\right)$.

Or, $x^2 + y^2 - c^2 = c^2 \left(\sinh^2\alpha - \sin^2\beta\right)$.

Let

$$x^2 + y^2 - c^2 = c^2(\sinh^2\alpha - \sin^2\beta) = f_2(x, y, c) \qquad (3.5.1)$$

And

$$x^2 + y^2 + c^2 = c^2(2 + \sinh^2\alpha - \sin^2\beta) = f_3(x, y, c) \qquad (3.5.2)$$

Then $f_3 + f_2 = 2(x^2 + y^2)$ and $f_3 - f_2 = 2\,c^2$ such that $[f_3]^2 - [f_2]^2 = [f_3 + f_2][f_3 - f_2] = 4\,c^2[x^2 + y^2]$.

Thus,

$$[f_2(x, y, c)]^2 = [f_3(x, y, c)]^2 - 4c^2[x^2 + y^2] \qquad (3.5.3)$$

Let

$$[f_2(x, y, c)]^2 + 4c^2y^2 = [f_3(x, y, c)]^2 - 4c^2x^2 = f_1(x, y, c) \qquad (3.5.4)$$

From equation (3.3.3), we have $c^2 = \frac{x^2}{\cosh^2\alpha} + \frac{y^2}{\sinh^2\alpha}$.

Or, $c^2 = [\frac{x^2}{1 + \sinh^2\alpha} + \frac{y^2}{\sinh^2\alpha}]$.

Or, $c^2 \sinh^2\alpha \left(1 + \sinh^2\alpha\right) = x^2\sinh^2\alpha + y^2\left(1 + \sinh^2\alpha\right)$.

Or, $c^2 \sinh^4\alpha + (c^2 - x^2 - y^2) \sinh^2\alpha - y^2 = 0$.

Or, $\sinh^2\alpha = -\frac{(c^2 - x^2 - y^2)}{2c^2} \pm \frac{\sqrt{[(c^2 - x^2 - y^2)^2 - 4c^2(-y^2)]}}{2c^2}$

$$= \frac{(x^2+y^2-c^2)}{2c^2} \pm \frac{\sqrt{[(x^2+y^2-c^2)^2+4c^2y^2]}}{2c^2}$$

$$= \frac{f_2}{2c^2} \pm \frac{\sqrt{[(f_2)^2+4c^2y^2]}}{2c^2} = \frac{f_2}{2c^2} \pm \frac{\sqrt{f_1}}{2c^2}$$

From equation (3.5.2), we note that f_3 is always greater than zero, since it is a sum of three squares and for an ellipse to exist c must be greater than zero. Thus, the lowest value of f_3 is $x^2 + y^2 + c^2 = c^2$ at the origin, where $x = y = 0$. Furthermore, on all points on the circumference of a circle of radius r, $x^2 + y^2 = r^2$ and $f_3 = x^2 + y^2 + c^2 = r^2 + c^2$. Hence, on all points on the circumference of the circle of radius c, $x^2 + y^2 = c^2$ and $f_3 = x^2 + y^2 + c^2 = 2\,c^2$

From equation (3.5.1), we note that f_2 can be negative. Thus, the lowest value of f_2 is $x^2 + y^2 - c^2 = -c^2$ at the origin, where $x = y = 0$. Furthermore, on all points on the circumference of a circle of radius r, $x^2 + y^2 = r^2$ and $f_2 = x^2 + y^2 - c^2 = r^2 - c^2$. Hence, on all points on the circumference of the circle of radius c, $x^2 + y^2 = c^2$ and $f_2 = x^2 + y^2 - c^2 = 0$. Inside this circle, f_2 is negative since $r < c$ and outside, it is positive since $r > c$.

Consequently, on all points on the circumference of a circle of radius r, $f_3 + f_2 = (r^2 + c^2) + (r^2 - c^2) = 2\,r^2$ and $f_3 - f_2 = (r^2 + c^2) - (r^2 - c^2) = 2\,c^2$ such that $[f_3]^2 - [f_2]^2 = [f_3 + f_2][f_3 - f_2] = 4\,c^2\,r^2$. Hence, at the origin, $[f_3]^2 = (c^2)^2 = [f_2]^2 = (-c^2)^2 = c^4$ such that $[f_3]^2 - [f_2]^2 = 0$. This is the lowest value of $[f_3]^2 - [f_2]^2$. Everywhere else, it is $4\,c^2[x^2 + y^2] > 0$.

From equation (3.5.4), we note that $f_1 = [f_2]^2 + 4c^2y^2$. Therefore, $f_1 = [f_2]^2$ on all points along x-axis where $y = 0$.

And, $f_1 = 4c^2y^2$ on all points on the circumference of the circle of radius c where $f_2 = 0$. Consequently, at the origin, $f_1 = [f_2]^2 = c^4$. As we move along the x-axis, away from the origin, the magnitude of f_2 decreases, although f_2 is negative, until $x = c$ or $x = -c$. Hence, f_1 decreases from c^4. At points $(-c, 0)$ and $(c, 0)$, $f_1 = f_2 = 0$. As we move along x-axis, beyond (-c,0) and (c,0), away from the origin, f_1 increases since f_2 increases. Everywhere else, where $y \neq 0$, $f_1 = [f_2]^2 + 4c^2y^2 > f_2^2$. Thus, $f_1 \geq f_2^2 \geq 0$.

At all points along y-axis, $x = 0$. Hence, $f_1 = [f_2]^2 + 4c^2y^2 = [x^2 + y^2 - c^2]^2 + 4c^2y^2 = [0 + y^2 - c^2]^2 + 4c^2y^2 = [y^2 + c^2]^2 = [f_3]^2$. Also, from equation (3.5.4), we note that $f_1 = [f_3]^2 - 4c^2x^2$. Therefore, $f_1 = [f_3]^2$ on all points along y-axis where $x = 0$. Everywhere else, where $x \neq 0$, $f_1 = [f_3]^2 - 4c^2x^2 < f_3^2$. Thus, $f_1 \leq f_3^2$.

We now conclude that $f_1 = 0$ at points $(-c, 0)$ and $(c, 0)$. Everywhere else, $f_1 > 0$. Therefore, $\sqrt{f_1}$ is real (not imaginary) everywhere.

And, using equation (3.5.4), we can also conclude that on all points above or below x-axis where $y \neq 0$, $\sqrt{f_1} = \sqrt{[f_2]^2 + 4c^2y^2} > f_2$ such that $f_2 - \sqrt{f_1} < 0$ even when $f_2 > 0$. However, $\sinh^2\alpha$ cannot be negative. Therefore, $\sinh^2\alpha = \frac{f_2}{2c^2} - \frac{\sqrt{f_1}}{2c^2}$ is not a possible solution. On the other hand, $f_2 + \sqrt{f_1} > 0$ even if $f_2 < 0$ (inside the circle $x^2 + y^2 = c^2$). Hence, $\sinh^2\alpha = \frac{f_2}{2c^2} + \frac{\sqrt{f_1}}{2c^2}$ is a possible solution.

On all points on x-axis, $y = 0$, and $\sqrt{f_1} = \sqrt{[f_2]^2} = \pm f_2$.

We examine three regions along x-axis. The first region is between $(-\infty, 0)$ and $(-c, 0)$ where $f_2 = x^2 + y^2 - c^2 =$

$x^2 - c^2 > 0$. Hence, $f_2 + \sqrt{f_1} = f_2 + f_2 = 2f_2 > 0$. Therefore, $\sinh^2\alpha > 0$. We know that (α, β) co-ordinates of $(-c, 0)$ are $(0, \pi)$. On the left side of this point $\alpha > 0$. We note that as we approach $(-c, 0)$ from the left side, α, f_2, and f_1 reduce and approach 0, but do not equal to 0. As a result, $\sinh\alpha$ approaches 0. And $\sinh^2\alpha = \frac{f_2}{2c^2} - \frac{\sqrt{f_1}}{2c^2} = 0$ is not a possible solution. At point $(-c, 0)$, $f_2 = x^2 + y^2 - c^2 = (-c)^2 - c^2 = 0$, $f_1 = [f_2]^2 = 0$, $\sinh\alpha = 0$, and $\alpha = 0$. Thus, within the region between $(-\infty, 0)$ and $(-c, 0)$ as well as at $(-c, 0)$, $\sinh^2\alpha = \frac{f_2}{2c^2} + \frac{\sqrt{f_1}}{2c^2}$ is a solution.

The second region is between $(-c, 0)$ and $(c, 0)$ where $f_2 = x^2 + y^2 - c^2 = x^2 - c^2 < 0$. However, at every point in this region, $f_1 = [f_2]^2$ is positive. And, $\sqrt{f_1} = \sqrt{[f_2]^2} = f_2$. Thus, f_2 and $\sqrt{f_1}$ are equal in magnitude. Hence, if both are negative then $\sinh^2\alpha$ will be negative which is not possible. Thus, $\sinh^2\alpha = \frac{f_2}{2c^2} - \frac{\sqrt{f_1}}{2c^2}$ is not a possible solution. Since we know f_2 is negative, $\sqrt{f_1}$ must be positive leading to $f_2 + \sqrt{f_1} = 0$, $\sinh^2\alpha = 0$, and $\alpha = 0$. The (α, β) co-ordinates of $(-c, 0)$, $(0, 0)$, and $(c, 0)$ respectively are $(0, \pi)$, $(0, \frac{\pi}{2})$, and $(0, 0)$. At point $(c, 0)$, $f_2 = x^2 + y^2 - c^2 = (c)^2 - c^2 = 0$, $f_1 = [f_2]^2 = 0$, and $\sinh\alpha = 0$. Thus, within the region between $(-c, 0)$ and $(c, 0)$ as well as at $(c, 0)$, $\sinh^2\alpha = \frac{f_2}{2c^2} + \frac{\sqrt{f_1}}{2c^2}$ is a solution.

The third region is between $(c, 0)$ and $(\infty, 0)$ where $f_2 = x^2 + y^2 - c^2 = x^2 - c^2 > 0$. Hence, $f_2 + \sqrt{f_1} = f_2 + f_2 = 2f_2 > 0$. Therefore, $\sinh^2\alpha > 0$. We know that (α, β) co-ordinates of $(c, 0)$ are $(0, 0)$. On the right side of this point $\alpha > 0$. We note that as we approach $(c, 0)$ from the right side, α, f_2, and f_1 reduce and approach 0, but do not equal to 0.

As a result, $\sinh\alpha$ approaches 0. And $\sinh^2\alpha = \frac{f_2}{2c^2} - \frac{\sqrt{f_1}}{2c^2} = 0$ is not a possible solution. Thus, within the region between $(c,0)$ and $(\infty,0)$, $\sinh^2\alpha = \frac{f_2}{2c^2} + \frac{\sqrt{f_1}}{2c^2}$ is a solution.

We conclude that $\sinh^2\alpha = \frac{f_2}{2c^2} + \frac{\sqrt{f_1}}{2c^2}$ is the only possible solution.

Consequently, $\cosh^2\alpha = 1 + \sinh^2\alpha = 1 + \frac{f_2 + \sqrt{f_1}}{2c^2}$

$= \frac{2c^2 + f_2 + \sqrt{f_1}}{2c^2} = \frac{2c^2 + x^2 + y^2 - c^2 + \sqrt{f_1}}{2c^2} = \frac{x^2 + y^2 + c^2 + \sqrt{f_1}}{2c^2} = \frac{f_3 + \sqrt{f_1}}{2c^2}.$

Hence, $\tanh^2\alpha = \frac{\sinh^2\alpha}{\cosh^2\alpha} = \frac{f_2 + \sqrt{f_1}}{f_3 + \sqrt{f_1}}$

And,

$$\tanh\alpha = \sqrt{\frac{f_2 + \sqrt{f_1}}{f_3 + \sqrt{f_1}}} \qquad (3.5.5)$$

where f_1, f_2, and f_3 are functions of x, y, and c.

Since α cannot be negative, we accept only positive values for $\tanh\alpha$.

3.6 α in terms of x, y, and β

From equations $(3.3.1)$ and $(3.3.2)$, we obtain

$\cosh^2\alpha = \frac{x^2}{c^2 \cos^2\beta}$ and $\sinh^2\alpha = \frac{y^2}{c^2 \sin^2\beta}$.

Hence, $\tanh^2\alpha = \frac{\sinh^2\alpha}{\cosh^2\alpha} = \frac{y^2}{c^2 \sin^2\beta} \frac{c^2 \cos^2\beta}{x^2} = \frac{y^2 \cot^2\beta}{x^2}.$

And,

$$\tanh\alpha = \sqrt{\frac{y^2 \cot^2\beta}{x^2}} = \frac{y \cot\beta}{x}. \qquad (3.6.1)$$

Since α cannot be negative, we accept only positive values for $\tanh\alpha$.

3.7 β in terms of x, y, and c

From equation (3.4.1), we have $c^2 = \frac{x^2}{\cos^2\beta} - \frac{y^2}{\sin^2\beta}$.

Or, $c^2 = \frac{x^2}{1-\sin^2\beta} - \frac{y^2}{\sin^2\beta}$

Or, $c^2 \sin^2\beta \left(1 - \sin^2\beta\right) = x^2\sin^2\beta - y^2 \left(1 - \sin^2\beta\right)$.

Or, $c^2 \sin^4\beta + \left(x^2 + y^2 - c^2\right) \sin^2\beta - y^2 = 0$.

Or, $\sin^2\beta = -\frac{(x^2+y^2-c^2)}{2c^2} \pm \frac{\sqrt{[(x^2+y^2-c^2)^2-4c^2(-y^2)]}}{2c^2}$

$\qquad = -\frac{(x^2+y^2-c^2)}{2c^2} \pm \frac{\sqrt{[(x^2+y^2-c^2)^2+4c^2y^2]}}{2c^2}$

$\qquad = -\frac{f_2}{2c^2} \pm \frac{\sqrt{[(f_2)^2+4c^2y^2]}}{2c^2} = -\frac{f_2}{2c^2} \pm \frac{\sqrt{f_1}}{2c^2}$

From equation (3.5.4), we know that on all points above or below x-axis where $y \neq 0$, $\sqrt{f_1} = \sqrt{[f_2]^2 + 4c^2y^2} > f_2$ such that $-f_2 - \sqrt{f_1} < 0$ even when $f_2 < 0$ (inside the circle of radius c) since the magnitude of $\sqrt{f_1}$ is more than the magnitude of f_2 when $y \neq 0$. Since $\sin^2\beta$ cannot be negative, $\sin^2\beta = -\frac{f_2}{2c^2} - \frac{\sqrt{f_1}}{2c^2}$ is not acceptable.

For all points on x-axis, $y = 0$ and the magnitude of $\sqrt{f_1}$ is same as the magnitude of f_2. Hence, when $f_2 > 0$ (in the region between $(-\infty, 0)$ and $(-c, 0)$ as well as in the region between $(c, 0)$ and $(\infty, 0)$), $-f_2 - \sqrt{f_1} = -2f_2$. Since $\sin^2\beta$ cannot be negative, $\sin^2\beta = -\frac{f_2}{2c^2} - \frac{\sqrt{f_1}}{2c^2}$ is not acceptable. When $f_2 < 0$ (in the region between $(-c, 0)$ and $(c, 0)$), $-f_2 - \sqrt{f_1} = 0$ yielding $\sin^2\beta = 0$ so that $\beta = 0$ or $\beta = \pi$. This situation is not acceptable since in this region β values lie between $\beta = 0$ and $\beta = \pi$ (or between $\beta = \pi$ and $\beta = 2\pi$). For example, at the origin $(0, 0)$, $\beta = \frac{\pi}{2}$.

At points $(-c, 0)$ and $(c, 0)$, $\sqrt{f_1} = f_2 = 0$ such that $-f_2 - \sqrt{f_1} = -f_2 + \sqrt{f_1} = 0$. Hence, it does not matter which solution we accept.

We now conclude that $\sin^2\beta = -\frac{f_2}{2c^2} - \frac{\sqrt{f_1}}{2c^2}$ is not acceptable.

Therefore, $\sin^2\beta = -\frac{f_2}{2c^2} + \frac{\sqrt{f_1}}{2c^2}$.

Consequently, $\cos^2\beta = 1 - \sin^2\beta = 1 - \frac{-f_2 + \sqrt{f_1}}{2c^2}$

$$= \frac{2c^2 + f_2 - \sqrt{f_1}}{2c^2} = \frac{2c^2 + x^2 + y^2 - c^2 - \sqrt{f_1}}{2c^2}$$

$$= \frac{x^2 + y^2 + c^2 - \sqrt{f_1}}{2c^2} = \frac{f_3 - \sqrt{f_1}}{2c^2}$$

Hence, $\tan^2\beta = \frac{-f_2 + \sqrt{f_1}}{f_3 - \sqrt{f_1}}$.

And,

$$\tan\beta = \sqrt{\frac{-f_2 + \sqrt{f_1}}{f_3 - \sqrt{f_1}}}. \qquad (3.7.1)$$

Since β cannot be negative (β values range from 0 to 2π), we accept only positive values for $\tan\beta$.

3.8 $\quad \beta$ in terms of x, y, and α

From equations (3.3.1) and (3.3.2), we obtain

$\cos^2\beta = \frac{x^2}{c^2\cosh^2\alpha}$ and $\sin^2\beta = \frac{y^2}{c^2\sinh^2\alpha}$.

Hence, $\tan^2\beta = \frac{\sin^2\beta}{\cos^2\beta} = \frac{y^2}{c^2\sinh^2\alpha}\frac{c^2\cosh^2\alpha}{x^2} = \frac{y^2\coth^2\alpha}{x^2}$.

And,

$$\tan\beta = \sqrt{\frac{y^2\coth^2\alpha}{x^2}} = \frac{y\coth\alpha}{x}. \qquad (3.8.1)$$

Since β cannot be negative (β values range from 0 to 2π), we accept only positive values for $\tan\beta$.

3.9   References

[1] Johannes Weertman, Stress and strain potentials for mode I and mode II cracks, *Mechanics of Materials*, 37:543-550, 2005.

[2] D. Maugis, Stresses and displacements around cracks and elliptical cavities : Exact solutions, *Engineering Fracture Mechanics*, Vol. 43, No, 2 pp. 217-255, 1992.

Chapter 4

Useful partial derivatives in the Elliptical co-ordinate system

We now develop expressions for useful partial derivatives.

4.1 Expressions for $\frac{\partial c}{\partial x}$ and $\frac{\partial c}{\partial y}$

From equation (3.3.3) in the previous chapter, we know

$$c = \sqrt{\frac{x^2}{\cosh^2\alpha} + \frac{y^2}{\sinh^2\alpha}}$$

Therefore,

$$\frac{\partial c}{\partial x} = \frac{2x}{2\cosh^2\alpha\sqrt{\frac{x^2}{\cosh^2\alpha} + \frac{y^2}{\sinh^2\alpha}}} = \frac{x}{\cosh^2\alpha\sqrt{\frac{x^2\sinh^2\alpha + y^2\cosh^2\alpha}{\sinh^2\alpha\,\cosh^2\alpha}}}$$

$$= \frac{x \sinh\alpha \cosh\alpha}{\cosh^2\alpha\sqrt{x^2 \sinh^2\alpha + y^2 \cosh^2\alpha}}$$

Thus,

$$\frac{\partial c}{\partial x} = \frac{x \tanh\alpha}{\sqrt{x^2 \sinh^2\alpha + y^2 \cosh^2\alpha}} \qquad (4.1.1)$$

Similarly,

$$\frac{\partial c}{\partial y} = \frac{2y}{2 \sinh^2\alpha\sqrt{\frac{x^2}{\cosh^2\alpha} + \frac{y^2}{\sinh^2\alpha}}} = \frac{y}{\sinh^2\alpha\sqrt{\frac{x^2 \sinh^2\alpha + y^2 \cosh^2\alpha}{\sinh^2\alpha \cosh^2\alpha}}}$$

$$= \frac{y \sinh\alpha \cosh\alpha}{\sinh^2\alpha\sqrt{x^2 \sinh^2\alpha + y^2 \cosh^2\alpha}}$$

Hence,

$$\frac{\partial c}{\partial y} = \frac{y \coth\alpha}{\sqrt{x^2 \sinh^2\alpha + y^2 \cosh^2\alpha}} \qquad (4.1.2)$$

4.2   Expressions for $\frac{\partial \alpha}{\partial x}$ and $\frac{\partial \alpha}{\partial y}$

From equations (3.5.1), (3.5.2), and (3.5.4), in the previous chapter, we know

$f_2(x, y, c) = x^2 + y^2 - c^2$,

$f_3(x, y, c) = x^2 + y^2 + c^2$, and

$f_1(x, y, c) = [f_2(x, y, c)]^2 + 4c^2 y^2$.

Therefore, $\frac{\partial f_2}{\partial x} = 2\,x$, $\frac{\partial f_3}{\partial x} = 2\,x$, and

$\frac{\partial f_1}{\partial x} = 2\,f_2\,\frac{\partial f_2}{\partial x} = 2\,f_2\,2\,x = 4\,x\,(x^2 + y^2 - c^2)$.

From equation (3.5.5) in the previous chapter, we also know $\tanh\alpha = \sqrt{\frac{f_2 + \sqrt{f_1}}{f_3 + \sqrt{f_1}}}$.

Hence, $\tanh^2\alpha = \frac{f_2+\sqrt{f_1}}{f_3+\sqrt{f_1}}$.

And $2\tanh\alpha\,\text{sech}^2\alpha\,\frac{\partial\alpha}{\partial x}$

$$= \frac{\frac{\partial(f_2+\sqrt{f_1})}{\partial x}(f_3+\sqrt{f_1})}{(f_3+\sqrt{f_1})^2} - \frac{\frac{\partial(f_3+\sqrt{f_1})}{\partial x}(f_2+\sqrt{f_1})}{(f_3+\sqrt{f_1})^2}$$

$$= \frac{[\frac{\partial(f_2)}{\partial x}+\frac{\partial(\sqrt{f_1})}{\partial x}](f_3+\sqrt{f_1})}{(f_3+\sqrt{f_1})^2} - \frac{[\frac{\partial(f_3)}{\partial x}+\frac{\partial(\sqrt{f_1})}{\partial x}](f_2+\sqrt{f_1})}{(f_3+\sqrt{f_1})^2}$$

$$= \frac{[2x+\frac{\partial(\sqrt{f_1})}{\partial x}](f_3+\sqrt{f_1})}{(f_3+\sqrt{f_1})^2} - \frac{[2x+\frac{\partial(\sqrt{f_1})}{\partial x}](f_2+\sqrt{f_1})}{(f_3+\sqrt{f_1})^2}$$

$$= [\frac{[2x+\frac{\partial(\sqrt{f_1})}{\partial x}]}{(f_3+\sqrt{f_1})^2}][(f_3+\sqrt{f_1})-(f_2+\sqrt{f_1})]$$

$$= [2x+\frac{\partial(\sqrt{f_1})}{\partial x}][\frac{1}{(f_3+\sqrt{f_1})^2}][f_3-f_2]$$

$$= [2x+\frac{\partial(f_1)}{2\sqrt{f_1}}][\frac{1}{(f_3+\sqrt{f_1})^2}][(x^2+y^2+c^2)-(x^2+y^2-c^2)]$$

$$= [2x+\frac{4xf_2}{2\sqrt{f_1}}][\frac{1}{(f_3+\sqrt{f_1})^2}][2c^2]$$

$$= 2x[1+\frac{f_2}{\sqrt{f_1}}][\frac{2c^2}{(f_3+\sqrt{f_1})^2}] = [\frac{f_2+\sqrt{f_1}}{\sqrt{f_1}}][\frac{4\,x\,c^2}{(f_3+\sqrt{f_1})^2}]$$

$$= [\frac{f_2+\sqrt{f_1}}{(f_3+\sqrt{f_1})}][\frac{4\,x\,c^2}{\sqrt{f_1}(f_3+\sqrt{f_1})}] = [\tanh^2\alpha][\frac{4\,x\,c^2}{\sqrt{f_1}(f_3+\sqrt{f_1})}].$$

Thus, $2\tanh\alpha\,\text{sech}^2\alpha\,\frac{\partial\alpha}{\partial x} = [\tanh^2\alpha][\frac{4\,x\,c^2}{\sqrt{f_1}(f_3+\sqrt{f_1})}]$.

Or, $\text{sech}^2\alpha\,\frac{\partial\alpha}{\partial x} = [\tanh\alpha][\frac{2\,x\,c^2}{\sqrt{f_1}(f_3+\sqrt{f_1})}]$.

Or, $\frac{\partial\alpha}{\partial x} = [\frac{2\tanh\alpha\,\cosh^2\alpha\,x\,c^2}{\sqrt{f_1}(f_3+\sqrt{f_1})}]$.

Hence,

$$\frac{\partial\alpha}{\partial x} = \frac{x\,c^2\,\sinh(2\alpha)}{\sqrt{f_1}(f_3+\sqrt{f_1})} \qquad (4.2.1)$$

Similarly, $\frac{\partial f_2}{\partial y} = 2\,y$, $\frac{\partial f_3}{\partial y} = 2\,y$, and

$\frac{\partial f_1}{\partial y} = 2\,f_2\,\frac{\partial f_2}{\partial y} + 8\,c^2\,y = 2\,f_2\,(2y) + 8\,c^2\,y = 4\,y\,(x^2 + y^2 - c^2) + 8\,c^2\,y.$

Also, $2\tanh\alpha\,\operatorname{sech}^2\alpha\,\frac{\partial \alpha}{\partial y}$

$$= \frac{\frac{\partial(f_2+\sqrt{f_1})}{\partial y}(f_3+\sqrt{f_1})}{(f_3+\sqrt{f_1})^2} - \frac{\frac{\partial(f_3+\sqrt{f_1})}{\partial y}(f_2+\sqrt{f_1})}{(f_3+\sqrt{f_1})^2}$$

$$= \frac{[\frac{\partial(f_2)}{\partial y}+\frac{\partial(\sqrt{f_1})}{\partial y}](f_3+\sqrt{f_1})}{(f_3+\sqrt{f_1})^2} - \frac{[\frac{\partial(f_3)}{\partial y}+\frac{\partial(\sqrt{f_1})}{\partial y}](f_2+\sqrt{f_1})}{(f_3+\sqrt{f_1})^2}$$

$$= \frac{[2y+\frac{\partial(\sqrt{f_1})}{\partial y}](f_3+\sqrt{f_1})}{(f_3+\sqrt{f_1})^2} - \frac{[2y+\frac{\partial(\sqrt{f_1})}{\partial y}](f_2+\sqrt{f_1})}{(f_3+\sqrt{f_1})^2}$$

$$= [\frac{[2y+\frac{\partial(\sqrt{f_1})}{\partial y}]}{(f_3+\sqrt{f_1})^2}][(f_3+\sqrt{f_1})-(f_2+\sqrt{f_1})]$$

$$= [2y+\frac{\partial(\sqrt{f_1})}{\partial y}][\frac{1}{(f_3+\sqrt{f_1})^2}][f_3-f_2]$$

$$= [2y+\frac{\frac{\partial(f_1)}{\partial y}}{2\sqrt{f_1}}][\frac{1}{(f_3+\sqrt{f_1})^2}][(x^2+y^2+c^2)-(x^2+y^2-c^2)]$$

$$= [2y+\frac{(4yf_2+8c^2y)}{2\sqrt{f_1}}][\frac{1}{(f_3+\sqrt{f_1})^2}][2c^2]$$

$$= 2y[1+\frac{f_2+2c^2}{\sqrt{f_1}}][\frac{2c^2}{(f_3+\sqrt{f_1})^2}] = [\frac{f_3+\sqrt{f_1}}{\sqrt{f_1}}][\frac{4\,y\,c^2}{(f_3+\sqrt{f_1})^2}]$$

$$= [\frac{f_2+\sqrt{f_1}}{(f_3+\sqrt{f_1})}][\frac{4\,y\,c^2}{\sqrt{f_1}(f_2+\sqrt{f_1})}] = [\tanh^2\alpha][\frac{4\,y\,c^2}{\sqrt{f_1}(f_2+\sqrt{f_1})}].$$

Thus, $2\tanh\alpha\,\operatorname{sech}^2\alpha\,\frac{\partial \alpha}{\partial y} = [\tanh^2\alpha][\frac{4\,y\,c^2}{\sqrt{f_1}(f_2+\sqrt{f_1})}].$

Or, $\operatorname{sech}^2\alpha\,\frac{\partial \alpha}{\partial y} = [\tanh\alpha][\frac{2\,y\,c^2}{\sqrt{f_1}(f_2+\sqrt{f_1})}].$

Or, $\frac{\partial \alpha}{\partial y} = [\frac{2\tanh\alpha\,\cosh^2\alpha\,y\,c^2}{\sqrt{f_1}(f_2+\sqrt{f_1})}].$

Hence,

$$\frac{\partial \alpha}{\partial y} = \frac{y\,c^2\,\sinh(2\alpha)}{\sqrt{f_1}(f_2+\sqrt{f_1})} \qquad (4.2.2)$$

Chapter 5

Displacement and deformation gradients in the Elliptical co-ordinate system

Consider deformation of an elliptical hole with starting parameters (α_s, c_s) and ending parameters (α_e, c_e) such that the starting semi-major axis is a_s and the starting semi-minor axis is b_s, whereas the ending semi-major axis is a_e and the ending semi-minor axis is b_e. The starting positions for points (x_s, y_s) on the predeformed ellipse are described by $x_s = c_s \cosh\alpha_s \cos\beta$ and $y_s = c_s \sinh\alpha_s \sin\beta$. The ending positions for points (x_e, y_e) on the deformed ellipse are described by $x_e = c_e \cosh\alpha_e \cos\beta$ and $y_e = c_e \sinh\alpha_e \sin\beta$.

The displacement of a point from (x_s, y_s) to (x_e, y_e) is u which has two components u_x in direction x and u_y in direction y.

5.1 Expressions for $\frac{\partial u_x}{\partial x_s}$ and $\frac{\partial x_e}{\partial x_s}$

The component of u in x direction is $u_x = x_e - x_s$ such that

$$\frac{\partial u_x}{\partial x_s} = \frac{\partial x_e}{\partial x_s} - 1. \tag{5.1.1}$$

Furthermore, u_x is made up of two parts : 1) due to change in only α and 2) due to change in only c. Thus,

$$u_x = (x_e - x_s)_{change\ in\ only\ \alpha} + (x_e - x_s)_{change\ in\ only\ c}$$

$$= (c_e \cosh\alpha_e \cos\beta - c_e \cosh\alpha_s \cos\beta)$$

$$+ (c_e \cosh\alpha_s \cos\beta - c_s \cosh\alpha_s \cos\beta). \tag{5.1.2}$$

Differentiating equation (5.1.2) with respect to x_s and using equation (5.1.1), as shown in Appendix A, we obtain

$$\frac{\partial x_e}{\partial x_s} = \frac{c_e \cos\beta \sinh\alpha_s \left[\frac{\partial(\alpha_s)}{\partial x_s}\right] + \cosh\alpha_s \cos\beta\left[\frac{\partial(c_s)}{\partial x_s}\right] - 1}{c_e \cos\beta \sinh\alpha_e \left[\frac{\partial(\alpha_e)}{\partial x_e}\right] + \cosh\alpha_s \cos\beta\left[\frac{\partial(c_e)}{\partial x_e}\right] - 1} \tag{5.1.3}$$

where $\frac{\partial(\alpha_s)}{\partial x_s} = \frac{x_s c_s^2 \sinh(2\alpha_s)}{\sqrt{f_1(x_s,y_s,c_s)}[f_3(x_s,y_s,c_s)+\sqrt{f_1(x_s,y_s,c_s)}]},$

$$\frac{\partial(\alpha_e)}{\partial x_e} = \frac{x_e\, c_e^2 \sinh(2\alpha_e)}{\sqrt{f_1(x_e,y_e,c_e)}\,[f_3(x_e,y_e,c_e)+\sqrt{f_1(x_e,y_e,c_e)}]},$$

$$\frac{\partial(c_s)}{\partial x_s} = \frac{x_s \tanh\alpha_s}{\sqrt{x_s^2 \sinh^2\alpha_s + y_s^2 \cosh^2\alpha_s}},\ \text{and}$$

$$\frac{\partial(c_e)}{\partial x_e} = \frac{x_e \tanh\alpha_e}{\sqrt{x_e^2 \sinh^2\alpha_e + y_e^2 \cosh^2\alpha_e}}.$$

As shown in Appendix A, at the tip,

$$\frac{\partial x_e}{\partial x_s} = \frac{a_e}{a_s} \ \text{ and } \ \frac{\partial u_x}{\partial x_s} = \frac{a_e - a_s}{a_s}. \tag{5.1.4}$$

And, at the top,

$$\frac{\partial x_e}{\partial x_s} = 1 \ \text{ and } \ \frac{\partial u_x}{\partial x_s} = 0. \tag{5.1.5}$$

5.2 Expressions for $\frac{\partial u_x}{\partial y_s}$ and $\frac{\partial x_e}{\partial y_s}$

The component of u in x direction is $u_x = x_e - x_s$ such that

$$\frac{\partial u_x}{\partial y_s} = \frac{\partial x_e}{\partial y_s} - \frac{\partial x_s}{\partial y_s} = \frac{\partial x_e}{\partial y_s} - 0 = \frac{\partial x_e}{\partial y_s}. \tag{5.2.1}$$

Differentiating equation (5.1.2) with respect to y_s and using equation (5.2.1), as shown in Appendix B, we obtain

$$\frac{\partial x_e}{\partial y_s} = \frac{c_e \cos\beta \sinh\alpha_s\, [\frac{\partial(\alpha_s)}{\partial y_s}] + \cosh\alpha_s \cos\beta[\frac{\partial(c_s)}{\partial y_s}]}{c_e \cos\beta \sinh\alpha_e\, [\frac{\partial(\alpha_e)}{\partial x_e}] + \cosh\alpha_s \cos\beta[\frac{\partial(c_e)}{\partial x_e}] - 1} \tag{5.2.2}$$

where $\dfrac{\partial(\alpha_s)}{\partial y_s} = \dfrac{y_s\,c_s^2\,\sinh(2\alpha_s)}{\sqrt{f_1(x_s,y_s,c_s)}\,[f_2(x_s,y_s,c_s)+\sqrt{f_1(x_s,y_s,c_s)}]}$,

$\dfrac{\partial(\alpha_e)}{\partial x_e} = \dfrac{x_e\,c_e^2\,\sinh(2\alpha_e)}{\sqrt{f_1(x_e,y_e,c_e)}\,[f_3(x_e,y_e,c_e)+\sqrt{f_1(x_e,y_e,c_e)}]}$,

$\dfrac{\partial(c_s)}{\partial y_s} = \dfrac{y_s\,\coth\alpha_s}{\sqrt{x_s^2\,\sinh^2\alpha_s+y_s^2\,\cosh^2\alpha_s}}$, and

$\dfrac{\partial(c_e)}{\partial x_e} = \dfrac{x_e\,\tanh\alpha_e}{\sqrt{x_e^2\,\sinh^2\alpha_e+y_e^2\,\cosh^2\alpha_e}}$.

As shown in Appendix B, at the tip,

$$\frac{\partial x_e}{\partial y_s} = 0 \text{ and } \frac{\partial u_x}{\partial y_s} = \frac{\partial x_e}{\partial y_s} = 0. \qquad (5.2.3)$$

And, at the top,

$$\frac{\partial x_e}{\partial y_s} = 0 \text{ and } \frac{\partial u_x}{\partial y_s} = \frac{\partial x_e}{\partial y_s} = 0. \qquad (5.2.4)$$

5.3 Expressions for $\dfrac{\partial u_y}{\partial x_s}$ and $\dfrac{\partial y_e}{\partial x_s}$

The component of u in y direction is $u_y = y_e - y_s$ such that

$$\frac{\partial u_y}{\partial x_s} = \frac{\partial y_e}{\partial x_s} - \frac{\partial y_s}{\partial x_s} = \frac{\partial y_e}{\partial x_s} - 0 = \frac{\partial y_e}{\partial x_s}. \qquad (5.3.1)$$

Furthermore, u_y is made up of two parts : 1) due to change in only α and 2) due to change in only c. Thus,

$u_y = (y_e - y_s)_{change\ in\ only\ \alpha} + (y_e - y_s)_{change\ in\ only\ c}$

54

$$= (c_e \sinh\alpha_e \sin\beta - c_e \sinh\alpha_s \sin\beta)$$

$$+ (c_e \sinh\alpha_s \sin\beta - c_s \sinh\alpha_s \sin\beta). \qquad (5.3.2)$$

Differentiating equation (5.3.2) with respect to x_s and using equation (5.3.1), as shown in Appendix C, we obtain

$$\frac{\partial y_e}{\partial x_s} = \frac{c_e \sin\beta \cosh\alpha_s \left[\frac{\partial(\alpha_s)}{\partial x_s}\right] + \sinh\alpha_s \sin\beta\left[\frac{\partial(c_s)}{\partial x_s}\right]}{\left[c_e \sin\beta \cosh\alpha_e \left[\frac{\partial(\alpha_e)}{\partial y_e}\right] + \sinh\alpha_s \sin\beta\left[\frac{\partial(c_e)}{\partial y_e}\right] - 1\right]}$$

$$(5.3.3)$$

where $\frac{\partial(\alpha_s)}{\partial x_s} = \frac{x_s\, c_s^2 \sinh(2\alpha_s)}{\sqrt{f_1(x_s,y_s,c_s)}[f_3(x_s,y_s,c_s)+\sqrt{f_1(x_s,y_s,c_s)}]}$,

$\frac{\partial(\alpha_e)}{\partial y_e} = \frac{y_e\, c_e^2 \sinh(2\alpha_e)}{\sqrt{f_1(x_e,y_e,c_e)}[f_2(x_e,y_e,c_e)+\sqrt{f_1(x_e,y_e,c_e)}]}$

$\frac{\partial(c_s)}{\partial x_s} = \frac{x_s \tanh\alpha_s}{\sqrt{x_s^2 \sinh^2\alpha_s+y_s^2 \cosh^2\alpha_s}}$, and

$\frac{\partial(c_e)}{\partial y_e} = \frac{y_e \coth\alpha_e}{\sqrt{x_e^2 \sinh^2\alpha_e+y_e^2 \cosh^2\alpha_e}}$.

As shown in Appendix C, at the tip,

$$\frac{\partial y_e}{\partial x_s} = 0 \text{ and } \frac{\partial u_y}{\partial x_s} = \frac{\partial y_e}{\partial x_s} = 0. \qquad (5.3.4)$$

And, at the top,

$$\frac{\partial y_e}{\partial x_s} = 0 \text{ and } \frac{\partial u_y}{\partial x_s} = \frac{\partial y_e}{\partial x_s} = 0. \qquad (5.3.5)$$

5.4 Expressions for $\frac{\partial u_y}{\partial y_s}$ and $\frac{\partial y_e}{\partial y_s}$

The component of u in y direction is $u_y = y_e - y_s$ such that

$$\frac{\partial u_y}{\partial y_s} = \frac{\partial y_e}{\partial y_s} - \frac{\partial y_s}{\partial y_s} = \frac{\partial y_e}{\partial y_s} - 1. \tag{5.4.1}$$

Differentiating equation (5.3.2) with respect to y_s and using equation (5.4.1), as shown in Appendix D, we obtain

$$\frac{\partial y_e}{\partial y_s} = \frac{c_e \sin\beta \cosh\alpha_s \left[\frac{\partial(\alpha_s)}{\partial y_s}\right] + \sinh\alpha_s \sin\beta\left[\frac{\partial(c_s)}{\partial y_s}\right] - 1}{\left[c_e \sin\beta \cosh\alpha_e \left[\frac{\partial(\alpha_e)}{\partial y_e}\right] + \sinh\alpha_s \sin\beta\left[\frac{\partial(c_e)}{\partial y_e}\right] - 1\right]} \tag{5.4.2}$$

where $\frac{\partial(\alpha_s)}{\partial y_s} = \dfrac{y_s \, c_s^2 \sinh(2\alpha_s)}{\sqrt{f_1(x_s,y_s,c_s)}[f_2(x_s,y_s,c_s)+\sqrt{f_1(x_s,y_s,c_s)}]}$,

$\frac{\partial(\alpha_e)}{\partial y_e} = \dfrac{y_e \, c_e^2 \sinh(2\alpha_e)}{\sqrt{f_1(x_e,y_e,c_e)}[f_2(x_e,y_e,c_e)+\sqrt{f_1(x_e,y_e,c_e)}]}$,

$\frac{\partial(c_s)}{\partial y_s} = \dfrac{y_s \coth\alpha_s}{\sqrt{x_s^2 \sinh^2\alpha_s + y_s^2 \cosh^2\alpha_s}}$, and

$\frac{\partial(c_e)}{\partial y_e} = \dfrac{y_e \coth\alpha_e}{\sqrt{x_e^2 \sinh^2\alpha_e + y_e^2 \cosh^2\alpha_e}}$.

As shown in Appendix D, at the tip,

$$\frac{\partial y_e}{\partial y_s} = 1 \text{ and } \frac{\partial u_y}{\partial y_s} = \frac{\partial y_e}{\partial y_s} - 1 = 0. \tag{5.4.3}$$

And, at the top,

$$\frac{\partial y_e}{\partial y_s} = \frac{b_e}{b_s} \text{ and } \frac{\partial u_y}{\partial y_s} = \frac{\partial y_e}{\partial y_s} - 1 = \frac{b_e - b_s}{b_s}. \tag{5.4.4}$$

Chapter 6

Stress-strain relations and boundary conditions

The stress analysis connected the applied stresses to the induced stress field and the deformed shape (described by $\frac{b_e}{a_e}$ or α_e). The strain-displacement relations connected the predeformed size and shape of the hole (described by α_s and c_s) to the deformed size and shape (described by α_e and c_e). The stress-strain relations connect material properties to the overall material response that decides the ultimate configuration.

6.1   Stress-strain relations

The stress-strain relations in two dimensions are

$$e_{xx} = \frac{1}{E}[\sigma_{xx} - \nu\sigma_{yy}],$$

$$e_{yy} = \frac{1}{E}[\sigma_{yy} - \nu\sigma_{xx}], \text{ and}$$

$$e_{xy} = \frac{1+\nu}{E}[\sigma_{xy}] = e_{yx} = \frac{1+\nu}{E}[\sigma_{yx}] \qquad (6.1.1)$$

where E is Young's Modulus and represents the slope of the linear segment of a stress-strain relation (for linear elastic materials) and ν is Poisson's ratio.

Thus, at every point in the plate, we now have developed

1) expressions (which contain a parameter α_e) for stress components (presented in the first book),

2) strain-displacement relations (which contain parameters α_e and c_e), and

3) stress-strain relations which link these two sets of expressions through e_{xx}, e_{yy}, and e_{xy} i.e. strain components.

As discussed in Section 1.1, we need any two of the four parameters (i.e. a_e, b_e, c_e, and α_e) to define the size and shape of the deformed elliptical hole. In order to determine these parameters, we now use boundary conditions.

6.2   Boundary conditions

Now, we use results from the stress analysis detailed in the first book mentioned above. These results show that, at the tip, there exists only vertical stress,

$\sigma_{yy} = [S_{yy}][1 + (\frac{2a_e}{b_e})] - [S_{xx}]$.

Hence, at the tip, $\sigma_{xx} = \sigma_{xy} = \sigma_{yx} = 0$.

Substituting these values for stress components in equations (6.1.1), we obtain

$e_{xx} = \frac{1}{E}[-\nu\sigma_{yy}], e_{yy} = \frac{1}{E}[\sigma_{yy}]$, and

$e_{xy} = \frac{1+\nu}{E}[0] = e_{yx} = 0$.

Thus, at the tip, we obtain the following relations for only normal strains,

$e_{xx} = \frac{-\nu}{E}[S_{yy}(1 + \frac{2a_e}{b_e}) - S_{xx}]$,

$e_{yy} = \frac{1}{E}[S_{yy}(1 + \frac{2a_e}{b_e}) - S_{xx}]$, and

$$e_{xy} = e_{yx} = 0. \tag{6.2.1}$$

Similarly, results from the previous stress analysis show that, at the top, there exists only horizontal stress,

$\sigma_{xx} = [S_{xx}][1 + (\frac{2b_e}{a_e})] - [S_{yy}]$.

Hence, at the top, $\sigma_{yy} = \sigma_{xy} = \sigma_{yx} = 0$.

Substituting these values for stress components in equations (6.1.1), we obtain

$e_{xx} = \frac{1}{E}[\sigma_{xx}], e_{yy} = \frac{1}{E}[-\nu\sigma_{xx}]$, and

$e_{xy} = \frac{1+\nu}{E}[0] = e_{yx} = 0$.

Thus, at the top, we obtain the following relations for only normal strains,

$e_{xx} = \frac{1}{E}[S_{xx}(1 + \frac{2b_e}{a_e}) - S_{yy}]$,

$$e_{yy} = \frac{-\nu}{E}[S_{xx}(1 + \frac{2b_e}{a_e}) - S_{yy}], \text{ and}$$

$$e_{xy} = e_{yx} = 0. \tag{6.2.2}$$

Chapter 7

Definitions of strain

I have presented below four definitions of strain. These definitions appear in many textbooks. However, books that deal with small strains only normally discuss the definition of engineering strain only.

7.1 Definitions of strain

We begin with the general definition which is strain

$$e = \frac{\lambda^n - 1}{n} \tag{7.1.1}$$

where λ is the stretch ratio, defined as

$$\lambda = \frac{final\ length}{original\ length} = \frac{l}{L} \tag{7.1.2}$$

in uniaxial tension where L is the original length and l is the final length.

Below, we consider four definitions of strain.

When $n = 1$, we obtain

$$e_{eng} = \frac{\lambda^n - 1}{n} = \lambda - 1 = \frac{l}{L} - 1 = \frac{l - L}{L}. \qquad (7.1.3)$$

This is the definition for **_engineering_** strain.

When $n = 2$, we obtain

$$e_{Green} = \frac{\lambda^2 - 1}{2} = \frac{1}{2}\left[\left(\frac{l}{L}\right)^2 - 1\right] = \frac{l^2 - L^2}{2L^2}. \qquad (7.1.4)$$

This is the definition for **_Green_** strain.

When $n = 0$, we obtain

$$e_{log} = \frac{\lambda^n - 1}{n} = \frac{e^{n\, ln\lambda} - 1}{n} \quad \text{(Here } e \text{ is not strain. It is the}$$

$$\text{base of natural logarithm.)}$$

$$= \frac{[1 + (n\, ln\lambda) + \frac{(n\, ln\lambda)^2}{2!} + ++] - 1}{n} \quad \text{(Taylor series)}$$

$$= \frac{[(n\, ln\lambda) + \frac{n^2(ln\lambda)^2}{2!} + \frac{n^3(ln\lambda)^3}{3!} + + + +]}{n}$$

$$= [(ln\lambda) + \frac{n(ln\lambda)^2}{2!} + \frac{n^2(ln\lambda)^3}{3!} + + + +].$$

When $n = 0$, only the first term remains. All other terms have n as a factor. Hence, they reduce to 0.

Thus, when $n = 0$,

$$e_{log} = (ln\lambda) \qquad (7.1.5)$$

This is the definition for **_logarithmic_** strain which is also defined as

$e_{log} = \int_L^l de = \int_L^l (\frac{dl}{l}) = (ln(\frac{l}{L})) = (ln\lambda)$.

When $n = -2$, we obtain

$e_{Almansi} = \dfrac{\lambda^n - 1}{n} = \dfrac{\lambda^{(-2)} - 1}{-2} = \dfrac{1}{2}\left[1 - \dfrac{1}{\lambda^2}\right]$

$= \frac{1}{2}[1 - \frac{1}{(\frac{l}{L})^2}] = \frac{1}{2}[1 - \frac{L^2}{l^2}]$.

Thus, when $n = -2$

$$e_{Almansi} = \frac{1}{2}\left[\frac{l^2 - L^2}{l^2}\right] \qquad (7.1.6)$$

This is the definition for **_Almansi_** strain.

I want to emphasize that so far we have made no assumption about the magnitude of strain. Furthermore, three of these four definitions are non-linear.

Chapter 8

Engineering strain

8.1 Applying boundary conditions

From equation (6.2.1), we know that at the tip,

$e_{xx} = \frac{-\nu}{E}[S_{yy}(1 + \frac{2a_e}{b_e}) - S_{xx}]$,

$e_{yy} = \frac{1}{E}[S_{yy}(1 + \frac{2a_e}{b_e}) - S_{xx}]$, and

$e_{xy} = e_{yx} = 0$.

Furthermore, from Chapter 2 and Chapter 5, we know that at the tip, the displacement is only in the horizontal direction. Therefore, at the tip, $\lambda = \frac{a_e}{a_s}$.

From equation (7.1.3), we know that $e_{eng} = \lambda - 1$. Hence, at the tip, $e_{xx} = \frac{a_e}{a_s} - 1$.

Therefore, $\frac{a_e}{a_s} - 1$

$$= \frac{-\nu}{E}[S_{yy}(1 + \frac{2a_e}{b_e}) - S_{xx}] = \frac{\nu}{E}[S_{xx} - S_{yy}(1 + \frac{2a_e}{b_e})]. \quad (8.1.1)$$

Similarly, from equation (6.2.2), we know that at the top,

$e_{xx} = \frac{1}{E}[S_{xx}(1 + \frac{2b_e}{a_e}) - S_{yy}]$,

$e_{yy} = \frac{-\nu}{E}[S_{xx}(1 + \frac{2b_e}{a_e}) - S_{yy}]$, and

$e_{xy} = e_{yx} = 0$.

Furthermore, from equation (5.1.5), we know that at the top, $\frac{\partial x_e}{\partial x_s} = 1$ and $\frac{\partial u_x}{\partial x_s} = 0$. Also, from Chapter 2 and Chapter 5, we know that at the top, the displacement is only in the vertical direction. Hence, at the top, $\lambda = \frac{b_e}{b_s}$.

From equation (7.1.3), we know that $e_{eng} = \lambda - 1$. Hence, at the top, $e_{yy} = \frac{b_e}{b_s} - 1$.

Therefore, $\frac{b_e}{b_s} - 1$

$$= \frac{-\nu}{E}[S_{xx}(1 + \frac{2b_e}{a_e}) - S_{yy}] = \frac{\nu}{E}[S_{yy} - S_{xx}(1 + \frac{2b_e}{a_e})]. \quad (8.1.2)$$

8.2 Evaluating a_e and b_e

From equation (8.1.2), as shown in Appendix E,

$$b_e = \frac{a_e b_s[1 + \frac{\nu}{E}S_{yy} - \frac{\nu}{E}S_{xx}]}{[a_e + 2b_s \frac{\nu}{E}S_{xx}]}. \quad (8.2.1)$$

Substituting this value for b_e in equation (8.1.1) and solving for a_e, as shown in Appendix E, we obtain

$$a_e = a_s b_s \left[\frac{[1 - (\frac{\nu}{E} S_{xx}) - (\frac{\nu}{E} S_{yy})][1 + (\frac{\nu}{E} S_{xx}) + (\frac{\nu}{E} S_{yy})]}{b_s(1 + \frac{\nu}{E} S_{yy} - \frac{\nu}{E} S_{xx}) + 2a_s(\frac{\nu}{E} S_{yy})} \right].$$

$$(8.2.2)$$

Expression for b_e in (8.2.1) still contains a_e. Hence, we substitute value for a_e from (8.2.2) in (8.2.1), as shown in Appendix E, to obtain

$$b_e = \frac{a_s b_s (1 - \frac{\nu}{E} S_{xx} - \frac{\nu}{E} S_{yy})(1 + \frac{\nu}{E} S_{xx} + \frac{\nu}{E} S_{yy})}{a_s[1 + (\frac{\nu}{E} S_{xx} - \frac{\nu}{E} S_{yy})] + 2(\frac{\nu}{E} S_{xx})[b_s]}. \quad (8.2.3)$$

We observe that when $a_s = 0$ or $b_s = 0$, both a_e and b_e vanish. Hence, a_s and b_s both must be greater than zero.

When $S_{xx} = 0$, these expressions simplify to

$$a_e = a_s b_s \left[\frac{[1 - (\frac{\nu}{E} S_{yy})][1 + (\frac{\nu}{E} S_{yy})]}{b_s(1 + \frac{\nu}{E} S_{yy}) + 2a_s(\frac{\nu}{E} S_{yy})} \right]$$

$$= b_s \left[\frac{[1 - (\frac{\nu}{E} S_{yy})^2]}{\frac{b_s}{a_s}(1 + \frac{\nu}{E} S_{yy}) + 2(\frac{\nu}{E} S_{yy})} \right]. \quad (8.2.4)$$

and $b_e = \dfrac{a_s b_s (1 - \frac{\nu}{E} S_{yy})(1 + \frac{\nu}{E} S_{yy})}{a_s[1 - \frac{\nu}{E} S_{yy}]}$

$$= b_s(1 + \frac{\nu}{E} S_{yy}). \quad (8.2.5)$$

Thus, when $S_{xx} = 0$, b_e is not affected by a_s.

Let us now consider two scenarios :

Scenario 1 : If $a_s = b_s = 1$ (i.e. a circular hole), $\nu = 0.3$, and $\frac{S_{yy}}{E} = 0.1$ then

$$a_e = b_s\left[\frac{1-[\frac{\nu}{E}S_{yy}]^2}{\frac{b_s}{a_s}(1+\frac{\nu}{E}S_{yy})+2(\frac{\nu}{E}S_{yy})}\right] = (1)\left[\frac{1-[(0.3)(0.1)]^2}{(1)[1+(0.3)(0.1)]+2[(0.3)(0.1)]}\right]$$

$$= \left[\frac{1-(0.03)^2}{(1+0.03)+2(0.03)}\right] = \left[\frac{1-0.0009}{1.03+0.06}\right] = \left[\frac{0.9991}{1.09}\right] = 0.9166$$

and $b_e = b_s(1 + \frac{\nu}{E}S_{yy}) = (1)[1+(0.3)(0.1)] = 1.03$

In this scenario, the predeformed circular hole becomes an elliptical hole with far field uniaxial stress along the major axis.

Hence, $a_s = 1.03$, $b_s = 0.9166$, $\nu = 0.3$, $\frac{S_{xx}}{E} = 0.1$, and $\frac{S_{yy}}{E} = 0$ such that

$$a_e = a_s b_s\left[\frac{[1-(\frac{\nu}{E}S_{xx})-(\frac{\nu}{E}S_{yy})][1+(\frac{\nu}{E}S_{xx})+(\frac{\nu}{E}S_{yy})]}{b_s(1+\frac{\nu}{E}S_{yy}-\frac{\nu}{E}S_{xx})+2a_s(\frac{\nu}{E}S_{yy})}\right]$$

$$= a_s b_s\left[\frac{[1-(\frac{\nu}{E}S_{xx})-(0)][1+(\frac{\nu}{E}S_{xx})+(0)]}{b_s(1+0-\frac{\nu}{E}S_{xx})+2a_s(0)}\right] = a_s b_s\left[\frac{[1-(\frac{\nu}{E}S_{xx})][1+(\frac{\nu}{E}S_{xx})]}{b_s(1-\frac{\nu}{E}S_{xx})}\right]$$

$$= a_s\left[1 + (\frac{\nu}{E}S_{xx})\right] = (1.03)\left[1+(0.3)(0.1)\right] = 1.0609$$

and $b_e = \dfrac{a_s b_s(1-\frac{\nu}{E}S_{xx}-\frac{\nu}{E}S_{yy})(1+\frac{\nu}{E}S_{xx}+\frac{\nu}{E}S_{yy})}{a_s[1+(\frac{\nu}{E}S_{xx}-\frac{\nu}{E}S_{yy})]+2(\frac{\nu}{E}S_{xx})[b_s]}$

$$= \frac{[1-(\frac{\nu}{E}S_{xx})^2]}{[1+\frac{\nu}{E}S_{xx}]\frac{1}{b_s}+2(\frac{\nu}{E}S_{xx})\frac{1}{a_s}} = \frac{1-[(0.3)(0.1)]^2}{[1+(0.3)(0.1)]\frac{1}{0.9166}+2(0.3)(0.1)\frac{1}{1.03}}$$

$$= \frac{1-(0.03)^2}{\frac{[1+(0.03)]}{0.9166}+\frac{2(0.03)}{1.03}} = \frac{1-0.0009}{\frac{1.03}{0.9166}+\frac{0.06}{1.03}} = \frac{(0.9991)(0.9166)(1.03)}{(1.03)(1.03)+(0.06)(0.9166)}$$

$$= \frac{0.9432483118}{1.0609+0.054996} = \frac{0.9432483118}{1.115896} = 0.845283$$

Thus, the final equilibrium value for a_e in the direction of the far field stress is 1.0609 and for b_e in the direction perpendicular to the far field stress is 0.845283.

Scenario 2 : If $a_s = 1.2$, $b_s = 1$ (i.e. an elliptical hole), $\nu = 0.3$, and $\frac{S_{yy}}{E} = 0.1$ then

$$a_e = b_s \left[\frac{1 - [\frac{\nu}{E}S_{yy}]^2}{\frac{b_s}{a_s}(1 + \frac{\nu}{E}S_{yy}) + 2(\frac{\nu}{E}S_{yy})} \right] = (1) \left[\frac{1 - [(0.3)(0.1)]^2}{\frac{1}{1.2}[1 + (0.3)(0.1)] + 2[(0.3)(0.1)]} \right]$$

$$= \left[\frac{1 - (0.03)^2}{\frac{1}{1.2}(1 + 0.03) + 2(0.03)} \right] = \left[\frac{(1.2)(1 - 0.0009)}{1.03 + (1.2)(0.06)} \right] = \left[\frac{(1.2)0.9991}{1.03 + 0.072} \right]$$

$$= \left[\frac{(1.2)0.9991}{1.102} \right] = \left[\frac{1.19892}{1.102} \right] = 1.0879$$

and $b_e = b_s(1 + \frac{\nu}{E}S_{yy}) = (1)[1 + (0.3)(0.1)] = 1.03$

Thus, the final equilibrium value for a_e in the direction perpendicular to the far field stress is 1.0879 and for b_e in the direction of the far field stress is 1.03.

In Scenario 1, a circular hole became an elliptical hole with its major axis along the direction of the far field stress. In Scenario 2, an elliptical hole remained an elliptical hole (both with major axis in the direction perpendicular to the far field stress) although $a_s/b_s = 1.2$ reduced to $a_e/b_e = 1.08795/1.03 = 1.0563$.

We conclude that a circular hole under uniaxial stress cannot remain a circular hole while an elliptical hole under uniaxial stress may remain an elliptical hole, although its focal length may change.

We now explore the situation when an elliptical hole becomes a circular hole under uniaxial stress (we treat $\underline{S_{xx} = 0}$). This situation requires that $a_e = b_e$. Using equations (8.2.4) and (8.2.5), we obtain,

$$a_e = b_s \left[\frac{[1 - (\frac{\nu}{E}S_{yy})^2]}{\frac{b_s}{a_s}(1 + \frac{\nu}{E}S_{yy}) + 2(\frac{\nu}{E}S_{yy})} \right] = b_e = b_s(1 + \frac{\nu}{E}S_{yy}).$$

Or, $[1 - (\frac{\nu}{E}S_{yy})^2] = [\frac{b_s}{a_s}(1 + \frac{\nu}{E}S_{yy}) + 2(\frac{\nu}{E}S_{yy})](1 + \frac{\nu}{E}S_{yy})$.

Or, $\frac{b_s}{a_s}(1 + \frac{\nu}{E}S_{yy})^2 = [1 - (\frac{\nu}{E}S_{yy})^2] - 2(\frac{\nu}{E}S_{yy})(1 + \frac{\nu}{E}S_{yy})$

$$= 1 - 3(\frac{\nu}{E}S_{yy})^2 - 2(\frac{\nu}{E}S_{yy}).$$

Thus, when $S_{xx} = 0$ if

$$\frac{b_s}{a_s} = \frac{1 - 3(\frac{\nu}{E}S_{yy})^2 - 2(\frac{\nu}{E}S_{yy})}{(1 + \frac{\nu}{E}S_{yy})^2} \qquad (8.2.6)$$

then a predeformed elliptical hole will become a circular hole after deformation.

$\nu \rightarrow$	0.1	0.2	0.3	0.4
$S_{yy}/E \downarrow$				
0.001	0.9996	0.9992	0.9988	0.998401
0.002	0.9992	0.998401	0.997601	0.996803
0.003	0.9988	0.997601	0.996403	0.995206
0.004	0.998401	0.996803	0.995206	0.99361
0.005	0.998001	0.996004	0.994009	0.992016
0.006	0.997601	0.995206	0.992813	0.990243
0.007	0.997202	0.994408	0.991618	0.988831
0.008	0.996803	0.99361	0.990423	0.987241
0.009	0.996403	0.992813	0.989229	0.985652
0.010	0.996004	0.992016	0.988036	0.984064

Table 8.2.1: b_s/a_s values for ten values of S_{yy}/E ranging from 0.001 to 0.01 and four values of ν 0.1, 0.2, 0.3, and 0.4 using equation (8.2.6).

Table 8.2.1 shows b_s/a_s values calculated using equation (8.2.6). As ν and/or S_{yy}/E increases b_s/a_s decreases. Once. a_s (or b_s) is known (or chosen), b_s (or a_s), a_e, and b_e can be calculated.

For example, when $S_{xx} = 0$, $\nu = 0.3$, and $\frac{S_{yy}}{E} = 0.1$,

$$\frac{b_s}{a_s} = \frac{1-3(\frac{\nu}{E}S_{yy})^2-2(\frac{\nu}{E}S_{yy})}{(1+\frac{\nu}{E}S_{yy})^2} = \frac{1-3[(0.3)(0.1)]^2-2(0.3)(0.1)}{[1+(0.3)(0.1)]^2}$$

$$= \frac{1-3(0.03)^2-2(0.03)}{[1+(0.03)]^2} = \frac{1-0.0027-0.06}{[1.03]^2} = 0.8835 = 1/1.1319.$$

Consequently, if we choose $a_s = 1$ and $b_s = 0.8835$ then

$$a_e = b_s\left[\frac{1-[\frac{\nu}{E}S_{yy}]^2}{\frac{b_s}{a_s}(1+\frac{\nu}{E}S_{yy})+2(\frac{\nu}{E}S_{yy})}\right]$$

$$= \frac{(0.8835)[1-[(0.3)(0.1)]^2]}{(0.8835)[1+(0.3)(0.1)]+2[(0.3)(0.1)]} = \frac{(0.8835)[1-(0.03)^2]}{(0.8835)[1+(0.03)]+2[(0.03)]}$$

$$= \frac{(0.8835)(1-0.0009)}{(0.8835)[1.03]+0.06} = \frac{(0.8835)(0.9991)}{0.910005+0.06} = \frac{0.88270485}{0.970005} = 0.91.$$

And, $b_e = b_s(1 + \frac{\nu}{E}S_{yy}) = (0.8835)[1 + (0.3)(0.1)] = (0.8835)(1.03) = 0.91$

such that the deformed circular hole will have a radius of 0.91.

However, if we choose $a_s = 1.1319$ and $b_s = 1$ then

$$a_e = b_s\left[\frac{1-[\frac{\nu}{E}S_{yy}]^2}{\frac{b_s}{a_s}(1+\frac{\nu}{E}S_{yy})+2(\frac{\nu}{E}S_{yy})}\right]$$

$$= (1)\left[\frac{1-[(0.3)(0.1)]^2}{\frac{1}{1.1319}[1+(0.3)(0.1)]+2[(0.3)(0.1)]}\right] = \left[\frac{1-(0.03)^2}{\frac{1}{1.1319}[1+(0.03)]+2[(0.03)]}\right]$$

$$= \left[\frac{(1.1319)(1-0.0009)}{[1.03]+(1.1319)0.06}\right] = \left[\frac{(1.1319)(0.9991)}{1.03+0.0679}\right] = \left[\frac{1.1309}{1.0979}\right] = 1.03.$$

And, $b_e = b_s(1 + \frac{\nu}{E}S_{yy}) = (1)[1 + (0.3)(0.1)] = 1.03$ such that the deformed circular hole will have a radius of 1.03.

Whether the deformed shape of the hole is circular or elliptical (the predeformed shape of the hole is known through a_s and b_s), using formulas developed in this section, a_e and b_e can be calculated, as illustrated above. With such calculated values for a_e and b_e, we can calculate values for α_e and c_e as well as stress components (using expressions developed in the first book) and gradients at any point (using expressions developed in Chapter 5 in this book).

8.3   Stress at the tip, point A

Analysis carried out in the first book showed that within the plate, the tip experiences the highest stress. Furthermore, at the tip, there exists only one stress component, $\sigma_{yy} = S_{yy}\left[1 + 2\frac{a_e}{b_e}\right] - S_{xx}$.

Substituting values for a_e and b_e from (8.2.2) and (8.2.3) in this expression, as shown in Appendix E, we obtain

$$\sigma_{yy} =$$

$$S_{yy} - S_{xx} + 2S_{yy}\left[\frac{a_s(1 + \frac{\nu}{E}S_{xx} - \frac{\nu}{E}S_{yy}) + 2b_s(\frac{\nu}{E}S_{xx})}{b_s(1 + \frac{\nu}{E}S_{yy} - \frac{\nu}{E}S_{xx}) + 2a_s(\frac{\nu}{E}S_{yy})}\right].$$

$$(8.3.1)$$

When $\underline{S_{xx} = 0}$ the above expression simplifies to

$$\sigma_{yy} = S_{yy} + 2S_{yy}\left[\frac{a_s(1 - \frac{\nu}{E}S_{yy})}{b_s(1 + \frac{\nu}{E}S_{yy}) + 2a_s(\frac{\nu}{E}S_{yy})}\right]$$

$$= S_{yy} \left[\frac{b_s(1 + \frac{\nu}{E}S_{yy}) + 2a_s}{b_s(1 + \frac{\nu}{E}S_{yy}) + 2a_s(\frac{\nu}{E}S_{yy})} \right]$$

Consequently, if the far field applied stress is only in the direction perpendicular to the major axis then the stress concentration, at the tip, is

$$\frac{\sigma_{yy}^{tip}}{S_{yy}} = \left[\frac{1 + (1 + \frac{\nu}{E}S_{yy})\frac{b_s}{2a_s}}{(\frac{\nu}{E}S_{yy}) + \frac{b_s}{2a_s}(1 + \frac{\nu}{E}S_{yy})} \right]. \qquad (8.3.2)$$

8.4 Comparison with Singh, Glinka, and Dubey (1994) paper

In the first book, I had mentioned in Chapter 9,

"(1) It will be shown that the Singh, Glinka, and Dubey (1994) [1] *finding is a special case of the general solution developed in Phase 2.*

(2) Two material properties, Young's modulus and Poisson's ratio affect the final dimensions (not just the Young's modulus)."

Accordingly, we now compare the expression (8.3.2) with a similar expression developed by Singh, Glinka, and Dubey (see expression (37) on page 485 of their paper)

$$\frac{\sigma_{22}^A}{S} = \frac{1 + [1 + (S/E)]\frac{b_i}{2a_i}}{S/E + (1 + (S/E))\frac{b_i}{2a_i}}. \qquad (8.4.1)$$

---

[1] M. Singh, G. Glinka, and R. Dubey
Notch and crack analysis as a moving boundary problem.
Engineering Fracture Mechanics, 47(4):479–492, 1994.

We note that if we let $\nu = 1$, in (8.3.2), we obtain (8.4.1). Thus, the expression developed by Singh, Glinka, and Dubey is a special case of the general solution developed here. Consequently, the expression developed by Singh, Glinka, and Dubey is applicable only to materials with Poisson's ratio equal to 1.

Finally, we note that expressions (8.2.2), (8.2.3), (8.3.1), and (8.3.2) all contain E and ν whereas expression (8.4.1) does not contain ν.

We observe that, although when $b_s = 0$ both the expressions (8.2.2) and (8.2.3) show that a_e and b_e also reduce to zero, if we let $b_s = 0$ in the expression (8.3.2), it reduces to $\frac{\sigma_{yy}^{tip}}{S_{yy}} = \left[\frac{1+0}{(\frac{\nu}{E} S_{yy}) + 0} \right] = \frac{E}{\nu} \frac{1}{S_{yy}}$. Or, when $S_{yy} > 0$ and $b_s = 0$

$$\sigma_{yy}^{tip} = \frac{E}{\nu}. \tag{8.4.2}$$

Similarly, if we let $b_i = 0$ in the expression (8.4.1), it reduces to $\frac{\sigma_{22}^{A}}{S} = \frac{1+0}{S/E + 0} = \frac{E}{S}$. Or, when $S > 0$ and $b_i = 0$

$$\sigma_{22}^{A} = E. \tag{8.4.3}$$

Hence, Singh, Glinka, and Dubey (see expression (44) and the subsequent paragraph on page 486 of their paper) stated, *"It is interesting to note that, contrary to the classical solution, the stress at the crack tip is finite. Moreover, according to eq. (44), the stress in the crack tip remains constant and*

*equal to the modulus of elasticity, E, regardless of the applied
load S and the initial length a_i."*

Furthermore, if we let $\frac{S}{E} = 1$ in (8.4.1) then

$$\frac{\sigma_{22}^A}{S} = \frac{1 + [1 + (S/E)]\frac{b_i}{2a_i}}{(S/E) + [1 + (S/E)]\frac{b_i}{2a_i}} = \frac{1 + [1 + 1]\frac{b_i}{2a_i}}{1 + [1 + 1]\frac{b_i}{2a_i}} = \frac{1 + \frac{b_i}{a_i}}{1 + \frac{b_i}{a_i}}$$

$$= 1.$$

Thus, when $\nu = 1$, the stress in the crack tip remains
constant and equal to the modulus of Elasticity, E, under
two circumstances, 1) when $b_i = 0$ regardless of the applied
load S and the initial length a_i and 2) when $S = E$ regardless
of the initial length a_i and height b_i.

We note that if we let $\nu = 1$, in (8.4.2), we obtain (8.4.3).
We obtain such results because, as discussed earlier, the expressions developed by Singh, Glinka, and Dubey are special
cases (where $\nu = 1$) of the general solution developed here.
Such expressions, consequently, have limited use.

On the other hand, as discussed in Appendix E, even if
two surfaces are closely touching each other, b_s exists, however small it may be. We can argue that if $b_s = 0$ then there
is no crack. Hence, $b_s > 0$. Thus, for sharp cracks where b_s
is much smaller than a_s, the tip stress approaches $\frac{E}{\nu}$.

8.5 Comparison with small deformation theory

I emphasize that so far we have imposed no restrictions on magnitudes of deformations. Hence, all expressions developed so far are applicable without any restrictions on magnitudes of deformations.

We now rewrite (8.3.2) as follows

$$\frac{\sigma_{yy}^{tip}}{E} = \frac{S_{yy}}{E}\left[\frac{1 + (1 + \nu\frac{S_{yy}}{E})\frac{1}{2}(\frac{b_s}{a_s})}{(\nu\frac{S_{yy}}{E}) + (1 + \nu\frac{S_{yy}}{E})\frac{1}{2}(\frac{b_s}{a_s})}\right]. \tag{8.5.1}$$

Using this expression, σ_{yy}^{tip}/E values when $\nu = 0.3$ are calculated for ten values of S_{yy}/E ranging from 0.1 to 1.0 and seven values of b_s/a_s, as shown in Table 8.5.1. It can be readily seen that as b_s/a_s approaches zero, σ_{yy}^{tip}/E approaches $1/\nu = 1/0.3 = 3.\bar{3}$.

The small deformation theory assumes that the deformations are small enough to ignore. Hence, based on the original dimensions, it provides us a simpler expression for $\frac{\sigma_{yy}^{tip}}{E}$ as follows :

$$\frac{\sigma_{yy}^{tip}}{E} = \frac{S_{yy}}{E}\left[1 + 2\left(\frac{a_s}{b_s}\right)\right]. \tag{8.5.2}$$

We note that the expression (8.5.1) contains ν whereas the expression (8.5.2) does not contain ν.

$b_s/a_s \rightarrow$ $S_{yy}/E \downarrow$	0.00000001	0.00001	0.0001	0.001	0.01	0.1	1
0.1	3.33333278	3.332778	3.32779	3.2788	2.8596	1.2902	0.2780
0.2	3.33333306	3.333057	3.33057	3.3059	3.0790	1.8637	0.5186
0.3	3.33333315	3.333150	3.33150	3.3151	3.1601	2.1893	0.7299
0.4	3.33333320	3.333196	3.33197	3.3197	3.2025	2.4000	0.9176
0.5	3.33333322	3.333225	3.33225	3.3225	3.2287	2.5482	1.0862
0.6	3.33333324	3.333244	3.33244	3.3244	3.2466	2.6586	1.2390
0.7	3.33333326	3.333257	3.33257	3.3258	3.2596	2.7444	1.3785
0.8	3.33333327	3.333268	3.33268	3.3268	3.2695	2.8132	1.5070
0.9	3.33333328	3.333276	3.33276	3.3276	3.2774	2.8700	1.6260
1.0	3.33333328	3.333283	3.33283	3.3283	3.2838	2.9178	1.7368

Table 8.5.1: σ_{yy}^{tip}/E values when $\nu = 0.3$ for ten values of S_{yy}/E ranging from 0.1 to 1.0 and seven values of b_s/a_s, as shown on the top line.

Now, consider the situation when $a_s = b_s$ (i.e. a circular hole before deformation). Hence, the expression (8.5.2) simplifies to

$$\frac{\sigma_{yy}^{tip}}{E} = 3\frac{S_{yy}}{E}. \qquad (8.5.3)$$

Whereas the expression (8.5.1) (which we developed without the assumption of small deformation) simplifies to

$$\frac{\sigma_{yy}^{tip}}{E} = \frac{S_{yy}}{E}\left[\frac{1 + (1 + \nu\frac{S_{yy}}{E})^{\frac{1}{2}}}{(\nu\frac{S_{yy}}{E}) + (1 + \nu\frac{S_{yy}}{E})^{\frac{1}{2}}}\right]. \qquad (8.5.4)$$

Now, consider the situation when $a_s = 100b_s$ (i.e. an elliptical hole before deformation). Hence, the expression (8.5.2) simplifies to

$$\frac{\sigma_{yy}^{tip}}{E} = 201\frac{S_{yy}}{E}. \qquad (8.5.5)$$

Whereas the expression (8.5.1) (which we developed without the assumption of small deformation) simplifies to

$$\frac{\sigma_{yy}^{tip}}{E} = \frac{S_{yy}}{E}\left[\frac{1 + (1 + \nu\frac{S_{yy}}{E})^{\frac{1}{2}}(0.01)}{(\nu\frac{S_{yy}}{E}) + (1 + \nu\frac{S_{yy}}{E})^{\frac{1}{2}}(0.01)}\right]. \qquad (8.5.6)$$

Using equations (8.5.3), (8.5.4), (8.5.5), and (8.5.6), σ_{yy}^{tip}/E values when $\nu = 0.3$ are calculated for ten values of S_{yy}/E ranging from 0.001 to 0.01 and two values of b_s/a_s, as shown in Table 8.5.2.

	Eq. 8.5.3	Eq. 8.5.4	Eq. 8.5.5	Eq. 8.5.6
$b_s/a_s \rightarrow$	1	1	0.01	0.01
$S_{yy}/E \downarrow$				
0.001	0.003	0.0030	0.201	0.1896
0.002	0.006	0.0060	0.402	0.3587
0.003	0.009	0.0090	0.603	0.5106
0.004	0.012	0.0120	0.804	0.6478
0.005	0.015	0.0149	1.005	0.7722
0.006	0.018	0.0179	1.206	0.8856
0.007	0.021	0.0209	1.407	0.9894
0.008	0.024	0.0238	1.608	1.0847
0.009	0.027	0.0268	1.809	1.1726
0.010	0.030	0.0298	2.010	1.2539

Table 8.5.2: σ_{yy}^{tip}/E values when $\nu = 0.3$ for ten values of S_{yy}/E ranging from 0.001 to 0.01 using equations (8.5.3), (8.5.4), (8.5.5), and (8.5.6).

We observe that the values for σ_{yy}^{tip}/E calculated using the small deformation theory are always higher except they are equal when $b_s = a_s$ and $S_{yy}/E = 0.001, 0.002, 0.003,$ and 0.004. The difference gets larger at lower values of b_s/a_s and higher values of S_{yy}/E.

Is this true for all materials? We note that (8.5.3) and (8.5.5) do not contain ν whereas (8.5.4) and (8.5.6) contain ν. To explore the effect of ν, the value for ν is changed from 0.3 to 0.1 in Table 8.5.2 to obtain Table 8.5.3 which is presented below. We find that at higher values of ν, the difference is larger.

	Eq. 8.5.3	Eq. 8.5.4	Eq. 8.5.5	Eq. 8.5.6
$b_s/a_s \rightarrow$	1	1	0.01	0.01
$S_{yy}/E \downarrow$				
0.001	0.003	0.0030	0.201	0.1970
0.002	0.006	0.0060	0.402	0.3865
0.003	0.009	0.0090	0.603	0.5687
0.004	0.012	0.0120	0.804	0.7442
0.005	0.015	0.0150	1.005	0.9132
0.006	0.018	0.0180	1.206	1.0762
0.007	0.021	0.0210	1.407	1.2335
0.008	0.024	0.0239	1.608	1.3853
0.009	0.027	0.0269	1.809	1.5319
0.010	0.030	0.0299	2.010	1.6736

Table 8.5.3: σ_{yy}^{tip}/E values when $\nu = 0.1$ for ten values of S_{yy}/E ranging from 0.001 to 0.01 using equations (8.5.3), (8.5.4), (8.5.5), and (8.5.6).

However small in magnitude, due to the presence of far field stresses, the deformations must occur. Hence, the values obtained by using the small deformation theory are not as reliable. On the other hand, they provide upper bounds to σ_{yy}^{tip}/E values.

Chapter 9

Green strain, logarithmic strain, and Almansi strain

We now move on to develop similar expressions using Green strain, logarithmic strain, and Almansi strain definitions.

9.1   Green strain

From Chapter 8, we know that at the tip,

$$e_{xx} = \frac{-\nu}{E}\left[S_{yy}(1 + \frac{2a_e}{b_e}) - S_{xx}\right] \text{ and } \lambda = \frac{a_e}{a_s}.$$

Also, at the top,

$$e_{yy} = \frac{-\nu}{E}\left[S_{xx}(1 + \frac{2b_e}{a_e}) - S_{yy}\right] \text{ and } \lambda = \frac{b_e}{b_s}.$$

Now, we use the definition of Green strain from (7.1.4)
$e_{Green} = \frac{\lambda^2 - 1}{2} = \frac{1}{2}[\lambda^2 - 1].$

Therefore, at the tip, $\frac{1}{2}[(\frac{a_e}{a_s})^2 - 1]$

$$= \frac{-\nu}{E}[S_{yy}(1 + \frac{2a_e}{b_e}) - S_{xx}] = \frac{\nu}{E}[S_{xx} - S_{yy}(1 + \frac{2a_e}{b_e})]. \quad (9.1.1)$$

Similarly, at the top, $\frac{1}{2}[(\frac{b_e}{b_s})^2 - 1]$

$$= \frac{-\nu}{E}[S_{xx}(1 + \frac{2b_e}{a_e}) - S_{yy}] = \frac{\nu}{E}[S_{yy} - S_{xx}(1 + \frac{2b_e}{a_e})]. \quad (9.1.2)$$

From equation (9.1.1) we know, at the tip,

$$\frac{1}{2}\left[\left(\frac{a_e}{a_s}\right)^2 - 1\right] = \frac{\nu}{E}\left[S_{xx} - S_{yy}\left(1 + \frac{2a_e}{b_e}\right)\right]. \quad (9.1.1)$$

Or, $\left[(\frac{a_e}{a_s})^2\right] = 1 + 2\frac{\nu}{E}\left[S_{xx} - S_{yy}(1 + \frac{2a_e}{b_e})\right].$

Or, $(a_e)^2 = (a_s)^2\left[1 + 2(\frac{\nu}{E})S_{xx} - 2(\frac{\nu}{E})S_{yy}(1 + \frac{2a_e}{b_e})\right]$

$$= (a_s)^2\left[1 + 2(\frac{\nu}{E})S_{xx} - 2(\frac{\nu}{E})S_{yy} - 2(\frac{\nu}{E})S_{yy}\frac{2a_e}{b_e}\right].$$

$$= (a_s)^2\left[1 + 2(\frac{\nu}{E})(S_{xx} - S_{yy}) - 4(\frac{\nu}{E})S_{yy}\frac{a_e}{b_e}\right].$$

Or, $(a_e)^2 + (4\frac{\nu}{E}S_{yy}\frac{a_s^2}{b_e})a_e - (a_s)^2\left[1 + 2(\frac{\nu}{E})(S_{xx} - S_{yy})\right] = 0.$

This is a quadratic equation in a_e with one unknown b_e. Hence,

$$a_e = \frac{-(4\frac{\nu}{E}S_{yy}\frac{a_s^2}{b_e})}{2} \pm \frac{\sqrt{(4\frac{\nu}{E}S_{yy}\frac{a_s^2}{b_e})^2 - 4[-(a_s)^2][1+2(\frac{\nu}{E})(S_{xx}-S_{yy})]}}{2}$$

$$= -(2\frac{\nu}{E}S_{yy}\frac{a_s^2}{b_e}) \pm \frac{2\sqrt{(2\frac{\nu}{E}S_{yy}\frac{a_s^2}{b_e})^2 + (a_s)^2[1+2(\frac{\nu}{E})(S_{xx}-S_{yy})]}}{2}$$

Since a_e cannot be negative,

$$a_e =$$

$$-\left(2\frac{\nu}{E}S_{yy}\frac{a_s^2}{b_e}\right) + a_s\sqrt{\left(2\frac{\nu}{E}S_{yy}\frac{a_s}{b_e}\right)^2 + 1 + 2\left(\frac{\nu}{E}\right)(S_{xx}-S_{yy})}. \tag{9.1.3}$$

Similarly, from equation (9.1.2) we know, at the top,

$$\frac{1}{2}\left[\left(\frac{b_e}{b_s}\right)^2 - 1\right] = \frac{\nu}{E}\left[S_{yy} - S_{xx}\left(1 + \frac{2b_e}{a_e}\right)\right]. \tag{9.1.2}$$

Or, $\left[(\frac{b_e}{b_s})^2\right] = 1 + 2\frac{\nu}{E}\left[S_{yy} - S_{xx}(1 + \frac{2b_e}{a_e})\right].$

Or, $(b_e)^2 = (b_s)^2\left[1 + 2(\frac{\nu}{E})S_{yy} - 2(\frac{\nu}{E})S_{xx}(1 + \frac{2b_e}{a_e})\right]$

$$= (b_s)^2\left[1 + 2(\frac{\nu}{E})S_{yy} - 2(\frac{\nu}{E})S_{xx} - 2(\frac{\nu}{E})S_{xx}\frac{2b_e}{a_e}\right].$$

$$= (b_s)^2\left[1 + 2(\frac{\nu}{E})(S_{yy} - S_{xx})\right] - 4(b_s)^2(\frac{\nu}{E})S_{xx}\frac{b_e}{a_e}.$$

Or, $(b_e)^2 + 4(b_s)^2(\frac{\nu}{E})S_{xx}\frac{b_e}{a_e} - (b_s)^2\left[1 + 2(\frac{\nu}{E})(S_{yy} - S_{xx})\right] = 0.$

Now, using value for a_e from equation (9.1.3), we obtain

$$4(b_s)^2(\frac{\nu}{E})S_{xx}\frac{b_e}{-\left(2\frac{\nu}{E}S_{yy}\frac{a_s^2}{b_e}\right) + a_s\sqrt{\left(2\frac{\nu}{E}S_{yy}\frac{a_s}{b_e}\right)^2 + 1 + 2\left(\frac{\nu}{E}\right)(S_{xx}-S_{yy})}}$$

$$+(b_e)^2 - (b_s)^2 \left[1 + 2(\tfrac{\nu}{E})(S_{yy} - S_{xx})\right] = 0.$$

Or,

$$(b_e)^2 \frac{4(b_s)^2(\tfrac{\nu}{E})S_{xx}}{-\left(2\tfrac{\nu}{E}S_{yy}a_s^2\right) + a_s\sqrt{\left(2\tfrac{\nu}{E}S_{yy}a_s\right)^2 + (b_e)^2\left[1+2\left(\tfrac{\nu}{E}\right)(S_{xx}-S_{yy})\right]}}$$

$$+ (b_e)^2 - (b_s)^2 \left[1 + 2(\frac{\nu}{E})(S_{yy} - S_{xx})\right] = 0. \qquad (9.1.4)$$

Equation (9.1.4) can be solved graphically or numerically to obtain b_e. Subsequently, we can use equation (9.1.3) to obtain a_e.

9.2   Logarithmic strain

From Chapter 8, we know that at the tip,

$$e_{xx} = \frac{-\nu}{E}\left[S_{yy}(1 + \tfrac{2a_e}{b_e}) - S_{xx}\right] \text{ and } \lambda = \frac{a_e}{a_s}.$$

Also, at the top,

$$e_{yy} = \frac{-\nu}{E}\left[S_{xx}(1 + \tfrac{2b_e}{a_e}) - S_{yy}\right] \text{ and } \lambda = \frac{b_e}{b_s}.$$

Now, we use the definition of logarithmic strain from (7.1.5) $e_{log} = ln(\lambda)$.

Therefore, at the tip, $ln(\frac{a_e}{a_s}) = \frac{-\nu}{E}[S_{yy}(1 + \tfrac{2a_e}{b_e}) - S_{xx}]$

Or, $\dfrac{a_e}{a_s} = e^{\frac{-\nu}{E}\left[S_{yy}(1+\tfrac{2a_e}{b_e})-S_{xx}\right]}$.

Or,

$$\frac{a_s}{a_e} = e^{\frac{\nu}{E}\left[S_{yy}(1+\frac{2a_e}{b_e})-S_{xx}\right]}. \qquad (9.2.1)$$

Similarly, at the top, $ln(\frac{b_e}{b_s}) = \frac{-\nu}{E}[S_{xx}(1 + \frac{2b_e}{a_e}) - S_{yy}]$

Or, $\frac{b_e}{b_s} = e^{\frac{-\nu}{E}\left[S_{xx}(1+\frac{2b_e}{a_e})-S_{yy}\right]}.$

Or,

$$\frac{b_s}{b_e} = e^{\frac{\nu}{E}\left[S_{xx}(1+\frac{2b_e}{a_e})-S_{yy}\right]}. \qquad (9.2.2)$$

Equations (9.2.1) and (9.2.2) can be solved graphically or numerically to obtain a_e and b_e.

In section 6.2, we noted that at the tip, there exists only one stress component which is $\sigma_{yy} = \left[S_{yy}(1 + \frac{2a_e}{b_e}) - S_{xx}\right]$ where σ_{yy} is Cauchy stress.

In the stress-strain relations, if we use Kirchoff stress in place of Cauchy stress then, at the tip, we obtain $e_{xx} = \frac{-\nu}{E}\tau_{yy}$ where $\tau_{yy} = J\sigma_{yy}$.

Furthermore, from chapter 5, we know gradients at the tip. Hence, $J = detF = \begin{bmatrix} \frac{\partial x_e}{\partial x_s} & \frac{\partial x_e}{\partial y_s} \\ \frac{\partial y_e}{\partial x_s} & \frac{\partial y_e}{\partial y_s} \end{bmatrix} = \begin{bmatrix} \frac{a_e}{a_s} & 0 \\ 0 & 1 \end{bmatrix} = \frac{a_e}{a_s}.$

Therefore, at the tip,

$e_{xx} = e_{log} = ln(\frac{a_e}{a_s}) = \frac{-\nu}{E}\tau_{yy} = \frac{-\nu}{E}J\sigma_{yy}$

$\qquad = \frac{-\nu}{E}\frac{a_e}{a_s}\left[S_{yy}(1 + \frac{2a_e}{b_e}) - S_{xx}\right].$

Or, $\dfrac{a_e}{a_s} = e^{\frac{-\nu}{E}\frac{a_e}{a_s}\left[S_{yy}(1+\frac{2a_e}{b_e})-S_{xx}\right]}$.

Or,

$$\dfrac{a_s}{a_e} = e^{\frac{\nu}{E}\frac{a_e}{a_s}\left[S_{yy}(1+\frac{2a_e}{b_e})-S_{xx}\right]}. \qquad (9.2.3)$$

Furthermore, from chapter 5, we know gradients at the top. Hence, $J = detF = \begin{bmatrix} \frac{\partial x_e}{\partial x_s} & \frac{\partial x_e}{\partial y_s} \\ \frac{\partial y_e}{\partial x_s} & \frac{\partial y_e}{\partial y_s} \end{bmatrix} = \begin{bmatrix} 1 & 0 \\ 0 & \frac{b_e}{b_s} \end{bmatrix} = \dfrac{b_e}{b_s}.$

Therefore, at the top,

$$e_{yy} = e_{log} = ln(\tfrac{b_e}{b_s}) = \tfrac{-\nu}{E}\tau_{xx} = \tfrac{-\nu}{E}J\sigma_{xx}$$

$$= \tfrac{-\nu}{E}\tfrac{b_e}{b_s}\left[S_{xx}(1+\tfrac{2b_e}{a_e}) - S_{yy}\right].$$

Or, $\dfrac{b_e}{b_s} = e^{\frac{-\nu}{E}\frac{b_e}{b_s}\left[S_{xx}(1+\frac{2b_e}{a_e})-S_{yy}\right]}$.

Or,

$$\dfrac{b_s}{b_e} = e^{\frac{\nu}{E}\frac{b_e}{b_s}\left[S_{xx}(1+\frac{2b_e}{a_e})-S_{yy}\right]}. \qquad (9.2.4)$$

Equations (9.2.3) and (9.2.4) can be solved graphically or numerically to obtain a_e and b_e.

9.3   Almansi strain

From Chapter 8, we know that at the tip,

$e_{xx} = \frac{-\nu}{E}\left[S_{yy}(1 + \frac{2a_e}{b_e}) - S_{xx}\right]$ and $\lambda = \frac{a_e}{a_s}$.

Also, at the top,

$e_{yy} = \frac{-\nu}{E}\left[S_{xx}(1 + \frac{2b_e}{a_e}) - S_{yy}\right]$ and $\lambda = \frac{b_e}{b_s}$.

Now, we use the definition of Almansi strain from (7.1.6)
$e_{Almansi} - \frac{1}{2}[1 - \frac{1}{\lambda^2}]$.

Therefore, at the tip, $\frac{1}{2}[1 - \frac{1}{(\frac{a_e}{a_s})^2}] = \frac{-\nu}{E}[S_{yy}(1 + \frac{2a_e}{b_e}) - S_{xx}]$.

Or, $[\frac{a_e^2 - a_s^2}{a_e^2}] = -2\frac{\nu}{E}[S_{yy} + S_{yy}\frac{2a_e}{b_e} - S_{xx}]$.

Or,

$$\left[4\frac{\nu}{E}S_{yy}\left(\frac{1}{b_e}\right)\right]a_e^3 + \left[1 + 2\frac{\nu}{E}(S_{yy} - S_{xx})\right]a_e^2 - a_s^2 = 0. \quad (9.3.1)$$

Similarly, at the top, $\frac{1}{2}[1 - \frac{1}{(\frac{b_e}{b_s})^2}] = \frac{-\nu}{E}[S_{xx}(1 + \frac{2b_e}{a_e}) - S_{yy}]$

Or, $[\frac{b_e^2 - b_s^2}{b_e^2}] = -2\frac{\nu}{E}[S_{xx} + S_{xx}\frac{2b_e}{a_e} - S_{yy}]$.

Or,

$$\left[4\frac{\nu}{E}S_{xx}\left(\frac{1}{a_e}\right)\right]b_e^3 + \left[1 + 2\frac{\nu}{E}(S_{xx} - S_{yy})\right]b_e^2 - b_s^2 = 0. \quad (9.3.2)$$

Equations (9.3.1) and (9.3.2) can be solved graphically or numerically to obtain a_e and b_e.

As in the section 9.2, if we use Kirchoff stress in place of Cauchy stress then, at the tip where $J = \frac{a_e}{a_s}$, we obtain

$\frac{1}{2}[1 - \frac{1}{(\frac{a_e}{a_s})^2}] = \frac{-\nu}{E}\frac{a_e}{a_s}[S_{yy}(1 + \frac{2a_e}{b_e}) - S_{xx}]$.

Or, $[\frac{a_e^2-a_s^2}{a_e^2}] = -2\frac{\nu}{E}\frac{a_e}{a_s}[S_{yy}(1 + \frac{2a_e}{b_e}) - S_{xx}]$. Therefore,

$$\left[4\frac{\nu}{E}S_{yy}\left(\frac{1}{b_e}\right)\right]a_e^4 + 2\frac{\nu}{E}\left[S_{yy} - S_{xx}\right]a_e^3 + a_s a_e^2 - a_s^3 = 0. \quad (9.3.3)$$

As in the section 9.2, if we use Kirchoff stress in place of Cauchy stress then, at the top where $J = \frac{b_e}{b_s}$, we obtain

$$\frac{1}{2}[1 - \frac{1}{(\frac{b_e}{b_s})^2}] = \frac{-\nu}{E}\frac{b_e}{b_s}[S_{xx}(1 + \frac{2b_e}{a_e}) - S_{yy}].$$

Or, $[\frac{b_e^2-b_s^2}{b_e^2}] = -2\frac{\nu}{E}\frac{b_e}{b_s}[S_{xx}(1 + \frac{2b_e}{a_e}) - S_{yy}]$. Therefore,

$$\left[4\frac{\nu}{E}S_{xx}\left(\frac{1}{a_e}\right)\right]b_e^4 + 2\frac{\nu}{E}\left[S_{xx} - S_{yy}\right]b_e^3 + b_s b_e^2 - b_s^3 = 0. \quad (9.3.4)$$

Equations (9.3.3) and (9.3.4) can be solved graphically or numerically to obtain a_e and b_e.

We now proceed to solve equations developed in this chapter to determine a_e and b_e.

Chapter 10

Procedures to compute a_e and b_e

As noted by Prof. Amit Singh (please see page 234), there exist sophisticated methods to solve non-linear equations. The intent here is to illustrate that it is possible to solve equations derived. The focus is on use of concepts related to the Elliptical co-ordinate system and the current configuration. Hence, the computational procedure used here is very simple and requires only background in using a spreadsheet such as Excel. For researchers interested in automating computations can refer to references similar to the two mentioned by Prof. Singh.

In Chapter 8, we discussed Engineering strain where we solved for a_e and b_e directly by substitution, since it was possible to isolate unknowns a_e and b_e. Furthermore, expressions developed there were linear.

However, for Green strain, logarithmic strain, and Almansi strain, equations we derived in Chapter 9 are non-linear.

10.1 Green strain

When $S_{xx} = 0$, or $S_{yy} = 0$, or when both S_{xx} and S_{yy} are non-zero, we need to solve

$$(b_e)^2 \frac{4(b_s)^2(\frac{\nu}{E})S_{xx}}{-\left(2\frac{\nu}{E}S_{yy}a_s^2\right) + a_s\sqrt{\left(2\frac{\nu}{E}S_{yy}a_s\right)^2 + (b_e)^2\left[1 + 2\left(\frac{\nu}{E}\right)(S_{xx} - S_{yy})\right]}}$$

$$+ (b_e)^2 - (b_s)^2\left[1 + 2(\frac{\nu}{E})(S_{yy} - S_{xx})\right] = 0. \qquad (9.1.4)$$

When $S_{xx} = 0$, equation (9.1.4) simplifies to

$(b_e)^2 - (b_s)^2\left[1 + 2(\frac{\nu}{E})(S_{yy})\right] = 0.$

Or,

$$(b_e) = (b_s)\sqrt{1 + 2\nu(\frac{S_{yy}}{E})}. \qquad (10.1.1)$$

For example, when $a_s = 1$, $b_s = 0.1$, $\nu = 0.3$, $\frac{S_{xx}}{E} = 0$, and $\frac{S_{yy}}{E} = 0.1$, as shown in Appendix F,

$b_e = 0.1029563014$ and consequently, $a_e = 0.5484325497$.

And when $S_{yy} = 0$, equation (9.1.4) simplifies to

$$(b_e)^2 \frac{4(b_s)^2(\frac{\nu}{E})S_{xx}}{a_s(b_e)\sqrt{\left[1+2(\frac{\nu}{E})(S_{xx})\right]}} + (b_e)^2 - (b_s)^2\left[1 - 2(\frac{\nu}{E})S_{xx}\right] = 0.$$

Or, as shown in Appendix F,

$$b_e = \frac{b_s\sqrt{4(\frac{b_s}{a_s})^2(\frac{\nu}{E})^2 S_{xx}^2 + \left[1 - 4(\frac{\nu}{E})^2 S_{xx}^2\right]} - 2b_s(\frac{b_s}{a_s})(\frac{\nu}{E})S_{xx}}{\sqrt{\left[1 + 2(\frac{\nu}{E})(S_{xx})\right]}}.$$

$$(10.1.2)$$

For example, when $a_s = 1$, $b_s = 0.1$, $\nu = 0.3$, $\frac{S_{xx}}{E} = 0.1$, and $\frac{S_{yy}}{E} = 0$, as shown in Appendix F,

$$b_e = 0.0968970714 \text{ and } a_e = 1.02956301409.$$

However, when both S_{xx} and S_{yy} are non-zero, it gets more complicated. Here, we substitute values for the known quantities in equation (9.1.4) and solve the simplified equation.

For example, when $a_s = 1$, $b_s = 0.1$, $\nu = 0.3$, $\frac{S_{xx}}{E} = 0.15$, and $\frac{S_{yy}}{E} = 0.1$, as shown in Appendix F, equation (9.1.4) simplifies to

$$\frac{(b_e)^2(0.0018)}{-0.06 + \sqrt{0.0036 + (b_e)^2 1.03}} + (b_e)^2 - 0.0097 = 0.$$

Let us label the left side of the above equation as $f(b_e)$.

We find b_e that satisfies this equation (i.e. $f(b_e) = 0$) by trial and error method. We notice that b_e^2 appears in the numerator in two terms while b_e^2 appears in the denominator of the first term under a square root. Hence, the left side of the equation increases as b_e increases. The first three rows of Table 10.1.1 provide a confirmation.

Next, we try $b_e = 0.1$ since our starting value is $b_s = 0.1$. We find that $f(b_e = 0.1) = 0.00061089016525$, still greater than 0. Hence, we try $b_e = 0.09$ to arrive at $f(b_e = 0.09) = -0.00130416382478$. Since $f(b_e)$ moved into negative territory, b_e must be between 0.09 and 0.1. When we try $b_e = 0.095$, halfway between 0.09 and 0.1, we obtain $f(b_e = 0.095) = -0.00037169221318$. If this level of accuracy is acceptable, then we stop here.

b_e	$f(b_e)$
1	0.99218154384845
2	3.99395358917313
3	8.99572666553153
0.1	0.00061089016525
0.09	-0.00130416382478
0.095	-0.00037169221318
0.098	0.00021184471220
0.097	0.00001532814242
0.096	-0.00017918419827
0.0969215694537	0.00000000000000

Table 10.1.1: $f(b_e)$ values for various (b_e) values.

Otherwise, we note that b_e must be between 0.095 and 0.01. We try a value between 0.095 and 0.01 such as $b_e = 0.098$ to obtain $f(b_e = 0.098) = 0.00021184471220$. Hence,

b_e must be between 0.095 and 0.098 and closer to 0.098 since 0.00021.. is closer to 0 than $-0.00037...$ We try $b_e = 0.097$ to obtain $f(b_e = 0.097) = 0.00001532814242$ and conclude that b_e must be between 0.095 and 0.097. When we try $b_e = 0.096$, we reach $f(b_e = 0.096) = -0.00017918419827$. Hence, b_e must be between 0.096 and 0.097. We continue to find a better value for b_e until the desired level of accuracy is reached. As shown in the last row of Table 10.1.1, $f(b_e = 0.0969215694537) = 0.00000000000000$.

The above procedure is not as cumbersome as it may seem. However, an algorithm can be developed to make it easier for the end user.

We observe that b_e appears as b_e^2 only. Hence, $f(b_e = -0.0969215694537) = 0.00000000000000$ is also a solution. However, b_e cannot be negative.

We accept $b_e = 0.0969215694537$ and move on to find a_e. As shown in Appendix F, using equation (9.1.3), we compute $a_e = 0.5697370473$.

Thus, using Green strain definition, when $a_s = 1$, $b_s = 0.1$, $\nu = 0.3$, $\frac{S_{xx}}{E} = 0.15$, and $\frac{S_{yy}}{E} = 0.1$, $a_s = 1$ reduces to $a_e = 0.5697370473$ and $b_s = 0.1$ reduces to $b_e = 0.0969215694$. To compare these results with the use of engineering strain, we recall equation (8.2.2) and equation (8.2.3). As shown in Appendix F, we obtain $a_e = 0.6273659306$ and $b_e = 0.0971069335$.

We conclude that both the models exhibit reductions in both major axis and minor axis. However, with the use of Green strain definition we arrive at larger reductions.

We now explore the situation when both S_{xx} and S_{yy} are compressive. As shown in Appendix F, when $a_s = 1$, $b_s = 0.1$, $\nu = 0.3$, $\frac{S_{xx}}{E} = -0.15$, and $\frac{S_{yy}}{E} = -0.1$, using Green strain definition, $a_s = 1$ increases to $a_e = 1.7336895082$ and $b_s = 0.1$ increases to $b_e = 0.10219819$. Once again, to compare these results with the use of engineering strain, we recall equation (8.2.2) and equation (8.2.3). As shown in Appendix F, we obtain $a_e = 2.3960843373$ and $b_e = 0.1018826844$.

10.2   Logarithmic strain

For logarithmic strain, we discussed two models in Chapter 9.

Part A : With Cauchy stress,

First, we discuss the model that involved Cauchy stress. We need to solve

$$\frac{a_s}{a_e} = e^{\frac{\nu}{E}\left[S_{yy}(1+\frac{2a_e}{b_e})-S_{xx}\right]}. \tag{9.2.1}$$

and

$$\frac{b_s}{b_e} = e^{\frac{\nu}{E}\left[S_{xx}(1+\frac{2b_e}{a_e})-S_{yy}\right]}. \tag{9.2.2}$$

When $\underline{S_{xx} = 0}$, equation (9.2.2) simplifies to

$$\frac{b_s}{b_e} = e^{\frac{\nu}{E}\left[S_{xx}(1+\frac{2b_e}{a_e})-S_{yy}\right]} = e^{\frac{\nu}{E}[-S_{yy}]}. \text{ Or, } \frac{b_e}{b_s} = e^{\frac{\nu}{E}[S_{yy}]}.$$

Or, $b_e = b_s e^{\frac{\nu}{E}[S_{yy}]}$ which is easy to work with.

For example, when $a_s = 1$, $b_s = 0.1$, $\nu = 0.3$, $\frac{S_{xx}}{E} = 0$ and $\frac{S_{yy}}{E} = 0.1$,

$$b_e = (0.1)e^{(0.3)(0.1)} = (0.1)e^{0.03} = (0.1)1.030454534$$

$$= 0.1030454534.$$

Hence, $\frac{a_s}{a_e} = e^{\frac{\nu}{E}[S_{yy}(1+\frac{2a_e}{b_e})-S_{xx}]} = e^{(0.3)[(0.1)(1+\frac{2a_e}{0.1030454534})-0]}$

$$= e^{[(0.03)+(\frac{0.06a_e}{0.1030454534})]} = e^{[0.03+0.5822673201a_e]}$$

Or, $a_s = 1 = a_e e^{[0.03+0.5822673201a_e]}$ which can be solved by trial and error method described in the previous section.

Table 10.2.1 is similar to the Table 10.1.1. The last row in Table 10.2.1 indicates that $a_e = 0.66058138149$.

a_e	$f(a_e)$
1	1.8446089801
0.5	0.6893449222
0.7	1.0842756714
0.6	0.8768097598
0.65	0.9779378510
0.66	0.9987817300
0.661	1.0008776444
0.66058138149	1.0000000000

Table 10.2.1: $f(a_e)$ values for various (a_e) values.

Thus, when $a_s = 1$, $b_s = 0.1$, $\nu = 0.3$, $\frac{S_{xx}}{E} = 0$, and

$\frac{S_{yy}}{E} = 0.1$, use of logarithmic strain definition and Cauchy stress yields $a_e = 0.66058138149$ and $b_e = 0.1030454534$.

When $\underline{S_{yy} = 0}$, equation (9.2.1) simplifies to

$$\frac{a_s}{a_e} = e^{\frac{\nu}{E}\left[S_{yy}(1+\frac{2a_e}{b_e})-S_{xx}\right]} = e^{\frac{\nu}{E}\left[-S_{xx}\right]}. \text{ Or, } \frac{a_e}{a_s} = e^{\frac{\nu}{E}[S_{xx}]}.$$

Or, $a_e = a_s e^{\frac{\nu}{E}[S_{xx}]}$.

When $a_s = 1$, $b_s = 0.1$, $\nu = 0.3$, $\frac{S_{xx}}{E} = 0.1$ and $\frac{S_{yy}}{E} = 0$,

$$a_e = a_s e^{\frac{\nu}{E}[S_{xx}]} = (1)e^{(0.3)(0.1)} = e^{0.03} = 1.030454534.$$

Hence, $\frac{b_s}{b_e} = e^{\frac{\nu}{E}[S_{xx}(1+\frac{2b_e}{a_e})-S_{yy}]} = e^{(0.3)[(0.1)(1+\frac{2b_e}{1.030454534})-0]}$

$$= e^{[0.03+\frac{0.06b_e}{1.030454534}]} = e^{[0.03+0.05822673201b_e]}$$

Or, $b_s = 0.1 = b_e e^{[0.03+0.05822673201b_e]}$ which can be solved by trial and error method described in the previous section.

b_e	$f(b_e)$
1	1.0922357388
2	2.3154421078
0.1	0.1036472036
0.09	0.0932281837
0.095	0.0984361812
0.096	0.0994781436
0.0965007963	0.1000000000

Table 10.2.2: $f(b_e)$ values for various (b_e) values.

Table 10.2.2 is similar to Table 10.1.1. The last row in Table 10.2.2 indicates that $b_e = 0.0965007963$.

Thus, when $a_s = 1$, $b_s = 0.1$, $\nu = 0.3$, $\frac{S_{xx}}{E} = 0.1$,

and $\frac{S_{yy}}{E} = 0$, use of logarithmic strain definition and Cauchy stress yields $a_e = 1.030454534$ and $b_e = 0.0965007963$.

However, when both S_{xx} and S_{yy} are non-zero, it gets more complicated. Here, we substitute values for the known quantities in equation (9.2.1) and equation (9.2.2) then solve the simplified equations.

For example, when $a_s = 1$, $b_s = 0.1$, $\nu = 0.3$, $\frac{S_{xx}}{E} = 0.15$, and $\frac{S_{yy}}{E} = 0.1$, equation (9.2.1) simplifies to

$$\frac{a_s}{a_e} = e^{\frac{\nu}{E}\left[S_{yy}(1+\frac{2a_e}{b_e})-S_{xx}\right]} = e^{(0.3)\left[(0.1)(1+\frac{2a_e}{b_e})-(0.15)\right]}$$

$$= e^{(0.3)\left[0.1+0.1\frac{2a_e}{b_e}-0.15\right]} = e^{(0.3)\left[\frac{0.2a_e}{b_e}-0.05\right]}.$$

Therefore,

$$e^{\left[\frac{0.06a_e}{b_e}-0.015\right]} - \frac{1}{a_e} = 0. \qquad (10.2.1)$$

Similarly, equation (9.2.2) simplifies to

$$\frac{b_s}{b_e} = e^{\frac{\nu}{E}\left[S_{xx}(1+\frac{2b_e}{a_e})-S_{yy}\right]} = e^{(0.3)\left[(0.15)(1+\frac{2b_e}{a_e})-(0.1)\right]}$$

$$= e^{(0.3)\left[0.15+0.15\frac{2b_e}{a_e}-0.1\right]} = e^{(0.3)\left[0.05+\frac{0.3b_e}{a_e}\right]}$$

Therefore,

$$e^{\left[0.015+\frac{0.09b_e}{a_e}\right]} - \frac{0.1}{b_e} = 0. \qquad (10.2.2)$$

We need to solve equation (10.2.1) and equation (10.2.2) each containing two unknowns, a_e and b_e. We employ the trial and error method described in detail in Appendix G to obtain five solutions that satisfy equation (10.2.1) and five solutions that satisfy equation (10.2.2). These ten solutions are plotted in Figure 10.2.1 showing two curves. The point of intersection of these two curves is the solution we are seeking.

Figure 10.2.1 indicates that the solution is in the vicinity of $a_e = 0.67$ and is somewhat less than $b_e = 0.1$.

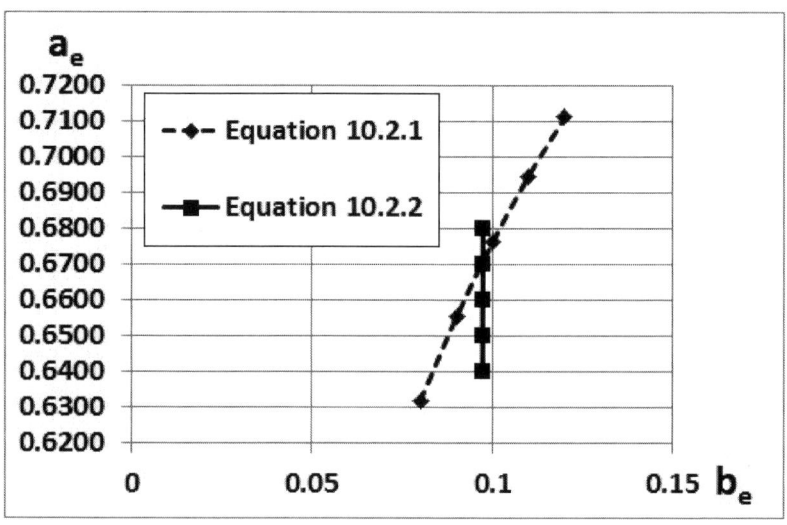

Figure 10.2.1: Plot of equation (10.2.1) and equation (10.2.2).

Figure 10.2.2 shows a close-up of that region. It can be readily seen that the solution is just above the point $a_e = 0.67$ and $b_e = 0.097232892$. We return to two matrices set up in Excel (as described in Appendix G) and, by trial and error method, we find that $a_e = 0.6709698413$ and $b_e = 0.097234704$ is the solution that satisfies both equation (10.2.1) and equation (10.2.2).

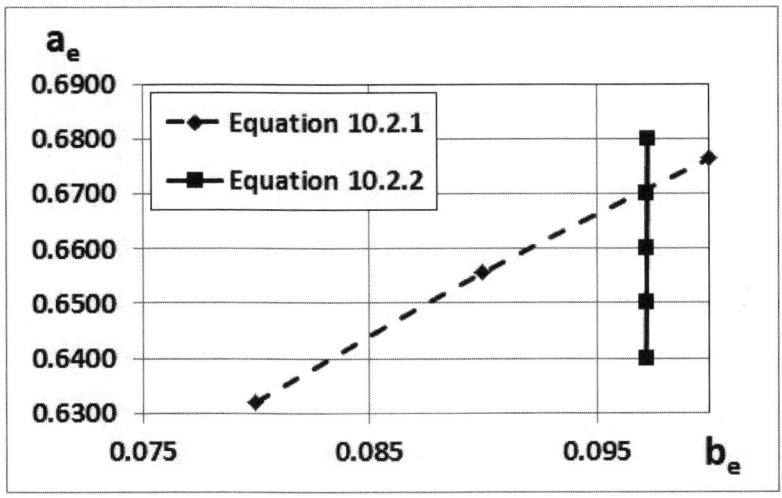

Figure 10.2.2: A close-up of Figure 10.2.1.

And, when $a_s = 1$, $b_s = 0.1$, $\nu = 0.3$, $\frac{S_{xx}}{E} = -0.15$, and $\frac{S_{yy}}{E} = -0.1$, equation (9.2.1) simplifies to

$$\frac{a_s}{a_e} = e^{\frac{\nu}{E}\left[S_{yy}(1+\frac{2a_e}{b_e})-S_{xx}\right]} = e^{(0.3)\left[(-0.1)(1+\frac{2a_e}{b_e})-(-0.15)\right]}$$

$$= e^{(0.3)\left[-0.1-0.1\frac{2a_e}{b_e}+0.15\right]} = e^{(0.3)\left[-\frac{0.2a_e}{b_e}+0.05\right]}.$$

Therefore, at the tip, $e^{\left[\frac{-0.06a_e}{b_e}+0.015\right]} - \frac{1}{a_e} = 0$.

Similarly, equation (9.2.2) simplifies to

$$\frac{b_s}{b_e} = e^{\frac{\nu}{E}\left[S_{xx}(1+\frac{2b_e}{a_e})-S_{yy}\right]} = e^{(0.3)\left[(-0.15)(1+\frac{2b_e}{a_e})-(-0.1)\right]}$$

$$= e^{(0.3)\left[-0.15-0.15\frac{2b_e}{a_e}+0.1\right]} = e^{(0.3)\left[-0.05-\frac{0.3b_e}{a_e}\right]}$$

Therefore, at the top, $e^{[-\frac{0.09b_e}{a_e}-0.015]} - \frac{0.1}{b_e} = 0$.

We need to solve $e^{[\frac{-0.06a_e}{b_e}+0.015]} - \frac{1}{a_e} = 0$ and

$e^{[-\frac{0.09b_e}{a_e}-0.015]} - \frac{0.1}{b_e} = 0$, each containing two unknowns, a_e and b_e. We employ the trial and error method described in detail in Appendix G. However, as shown in Figure 10.2.3, we find that there is no possible solution.

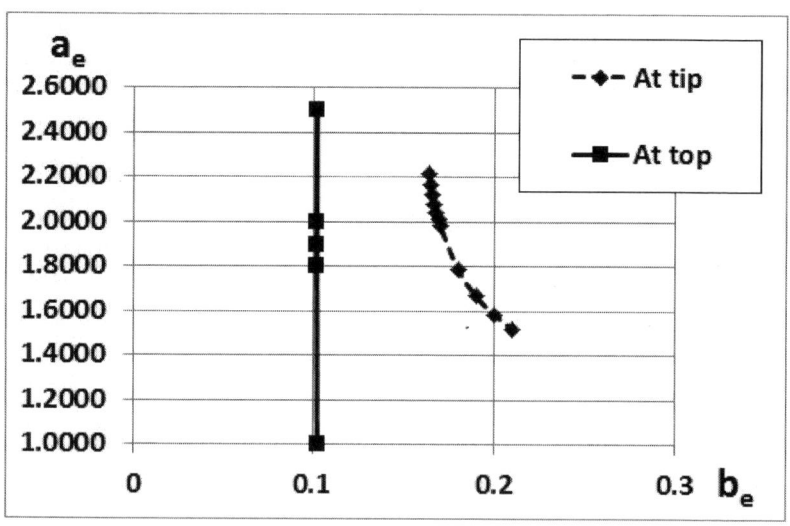

Figure 10.2.3: Plot showing two curves do not intersect.

This is a possible limitation on the use of models based on logarithmic strain definition.

Part B : With Kirchoff stress,

Now, we discuss the model that involved Kirchoff stress. We need to solve

$$\frac{a_s}{a_e} = e^{\frac{\nu}{E}\frac{a_e}{a_s}\left[S_{yy}(1+\frac{2a_e}{b_e})-S_{xx}\right]}. \tag{9.2.3}$$

and

$$\frac{b_s}{b_e} = e^{\frac{\nu}{E}\frac{b_e}{b_s}\left[S_{xx}(1+\frac{2b_e}{a_e})-S_{yy}\right]}. \tag{9.2.4}$$

When $S_{xx} = 0$, equation (9.2.4) simplifies to

$$\frac{b_s}{b_e} = e^{\frac{\nu}{E}\frac{b_e}{b_s}\left[S_{xx}(1+\frac{2b_e}{a_e})-S_{yy}\right]} = e^{\frac{\nu}{E}\frac{b_e}{b_s}\left[-S_{yy}\right]}.$$

Or,

$$b_e = b_s e^{\frac{\nu}{E}\frac{b_e}{b_s}[S_{yy}]} \tag{10.2.3}$$

which contains only one unknown b_e. Hence, it can be solved by trial and error method, as shown in section 10.1.

For example, when $a_s = 1$, $b_s = 0.1$, $\nu = 0.3$, $\frac{S_{xx}}{E} = 0$ and $\frac{S_{yy}}{E} = 0.1$,

we obtain $b_e = b_s e^{\frac{\nu}{E}\frac{b_e}{b_s}[S_{yy}]} = (0.1)e^{(0.3)\frac{b_e}{(0.1)}(0.1)} = (0.1)e^{(0.3)b_e}$.

By trial and error method, we find that the solution is $b_e = 0.1031426499$.

Hence, $\frac{a_s}{a_e} = e^{\frac{\nu}{E}\frac{a_e}{a_s}\left[S_{yy}(1+\frac{2a_e}{b_e})-S_{xx}\right]}$ yields

$\frac{1}{a_e} = e^{(0.3)\frac{a_e}{1}[(0.1)(1+\frac{2a_e}{0.1031426499})]}$.

Or, $a_e e^{[(0.03)a_e(1+\frac{2a_e}{0.1031426499})]} = 1$.

By trial and error method, we find that the solution is $a_e = 0.72236787845$. Thus, when we use Kirchoff stress, if $a_s = 1$, $b_s = 0.1$, $\nu = 0.3$, $\frac{S_{xx}}{E} = 0$ and $\frac{S_{yy}}{E} = 0.1$, then $a_e = 0.72236787845$ and $b_e = 0.1031426499$.

In comparison, when we used Cauchy stress, in part A, we found $a_e = 0.66058138149$ and $b_e = 0.1030454534$.

When $S_{yy} = 0$, equation (9.2.3) simplifies to

$$\frac{a_s}{a_e} = e^{\frac{\nu}{E}\frac{a_e}{a_s}\left[S_{yy}(1+\frac{2a_e}{b_e})-S_{xx}\right]} = e^{\frac{\nu}{E}\frac{a_e}{a_s}\left[-S_{xx}\right]}$$

Or,

$$a_e = a_s e^{\frac{\nu}{E}\frac{a_e}{a_s}[S_{xx}]} \qquad (10.2.4)$$

which can be solved by trial and error method, as shown in section 10.1.

For example, when $a_s = 1$, $b_s = 0.1$, $\nu = 0.3$, $\frac{S_{xx}}{E} = 0.1$ and $\frac{S_{yy}}{E} = 0$,

$a_e = a_s e^{\frac{\nu}{E}\frac{a_e}{a_s}[S_{xx}]} = (1)e^{(0.3)\frac{a_e}{1}[0.1]} = e^{(0.03)a_e}$

Or, $e^{(0.03)a_e} - a_e = 0$.

By trial and error method, we find that the solution is $a_e = 1.031426499$.

Hence, $\dfrac{b_s}{b_e} = e^{\frac{\nu}{E}\frac{b_e}{b_s}\left[S_{xx}(1+\frac{2b_e}{a_e})-S_{yy}\right]}$ yields

$$\dfrac{0.1}{b_e} = e^{(0.3)\frac{b_e}{0.1}\left[(0.1)(1+\frac{2b_e}{1.031426499})-0\right]} = e^{(0.3)b_e(1+\frac{2b_e}{1.031426499})}$$

Or, $b_e e^{[(0.3)b_e(1+\frac{2b_e}{1.031426499})]} = 0.1$

By trial and error method, we find that the solution is $b_e = 0.0966170115$. Thus, when we use Kirchoff stress, if $a_s = 1$, $b_s = 0.1$, $\nu = 0.3$, $\frac{S_{xx}}{E} = 0.1$ and $\frac{S_{yy}}{E} = 0$, then $a_e = 1.031426499$ and $b_e = 0.0966170115$.

In comparison, when we used Cauchy stress, in part A, we found $a_e = 1.030454534$ and $b_e = 0.0965007963$.

When both S_{xx} and S_{yy} are non-zero, we sub-
stitute values for the known quantities in equation (9.2.3) and equation (9.2.4) then solve the simplified equations.

For example, when $a_s = 1$, $b_s = 0.1$, $\nu = 0.3$, $\frac{S_{xx}}{E} = 0.15$, and $\frac{S_{yy}}{E} = 0.1$, equation (9.2.3) simplifies to

$$\dfrac{a_s}{a_e} = e^{\frac{\nu}{E}\frac{a_e}{a_s}\left[S_{yy}(1+\frac{2a_e}{b_e})-S_{xx}\right]} = e^{(0.3)\frac{a_e}{1}\left[(0.1)(1+\frac{2a_e}{b_e})-(0.15)\right]}$$

$$= e^{(0.3)a_e\left[(0.1)+(0.1)\frac{2a_e}{b_e}-(0.15)\right]} = e^{[(0.3)a_e(\frac{0.2a_e}{b_e}-0.05)]}$$

Therefore,

$$e^{[(0.3)a_e(\frac{0.2a_e}{b_e}-0.05)]} - \dfrac{1}{a_e} = 0. \qquad (10.2.5)$$

Similarly, equation (9.2.4) simplifies to

$$\frac{b_s}{b_e} = e^{\frac{\nu}{E}\frac{b_e}{b_s}\left[S_{xx}(1+\frac{2b_e}{a_e})-S_{yy}\right]} = e^{(0.3)\frac{b_e}{0.1}\left[(0.15)(1+\frac{2b_e}{a_e})-(0.1)\right]}$$

$$= e^{(3)b_e\left[((0.15)+(0.15)\frac{2b_e}{a_e})-(0.1)\right]} = e^{[(3)b_e(0.05+\frac{0.3b_e}{a_e})]}$$

Therefore,

$$e^{[(3)b_e(0.05+\frac{0.3b_e}{a_e})]} - \frac{0.1}{b_e} = 0. \tag{10.2.6}$$

We need to solve equation (10.2.5) and equation (10.2.6) each containing two unknowns, a_e and b_e. We employ the trial and error method described in Appendix G. Table 10.2.3 (similar to the condensed version of Table G.0.1) shows five solutions to equation (10.2.5) where b_e values are selected as 0.08, 0.09, 0.10, 0.11, and 0.12, while a_e values are found by trial and error so that equation (10.2.5) is satisfied.

a_e	b_e	$Equation - 10.2.5$
0.69986805685	0.08	0.0000000000
0.71730332083	0.09	0.0000000000
0.73265507715	0.10	0.0000000000
0.74631109750	0.11	0.0000000000
0.75856249439	0.12	0.0000000000

Table 10.2.3: Five solutions to equation (10.2.5).

Similarly, Table 10.2.4 (similar to the condensed version of Table G.0.2) shows five solutions to equation (10.2.6) where a_e values are selected as 0.70, 0.71, 0.72, 0.73, and

0.74, while b_e values are found by trial and error so that equation (10.2.6) is satisfied.

a_e	b_e	$Equation - 10.2.6$
0.70	0.09735657886	0.0000000000
0.71	0.09737266877	0.0000000000
0.72	0.09738832434	0.0000000000
0.73	0.09740356293	0.0000000000
0.74	0.097418400975	0.0000000000

Table 10.2.4: Five solutions to equation (10.2.6).

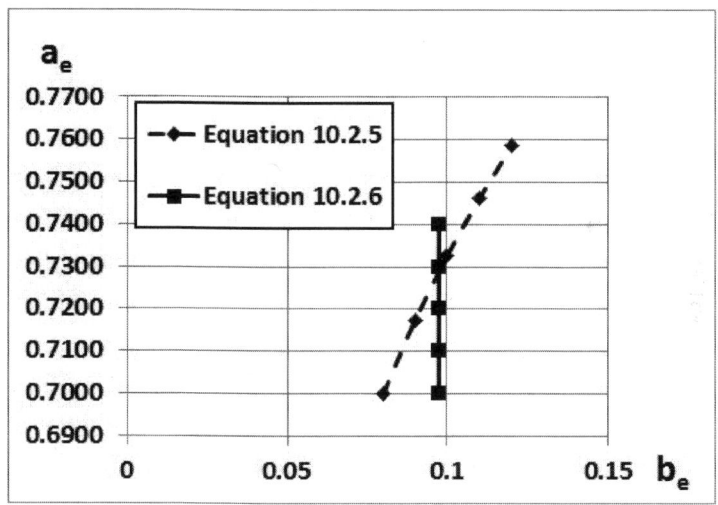

Figure 10.2.4: Plot of equation (10.2.5) and equation (10.2.6).

These ten solutions are plotted in Figure 10.2.4 which shows that the solution is in the vicinity of the fourth solution in Table 10.2.4. Figure 10.2.5 shows a close-up of that region. It can be readily seen that the solution is just below the point $a_e = 0.73$ and $b_e = 0.09740356293$.

By trial and error method, we find that $a_e = 0.7288431351$ and $b_e = 0.09740182083$ is the solution that satisfies both equation (10.2.5) and equation (10.2.6).

In comparison, when we used Cauchy stress, in part A, we found $a_e = 0.6709698413$ and $b_e = 0.097234704$.

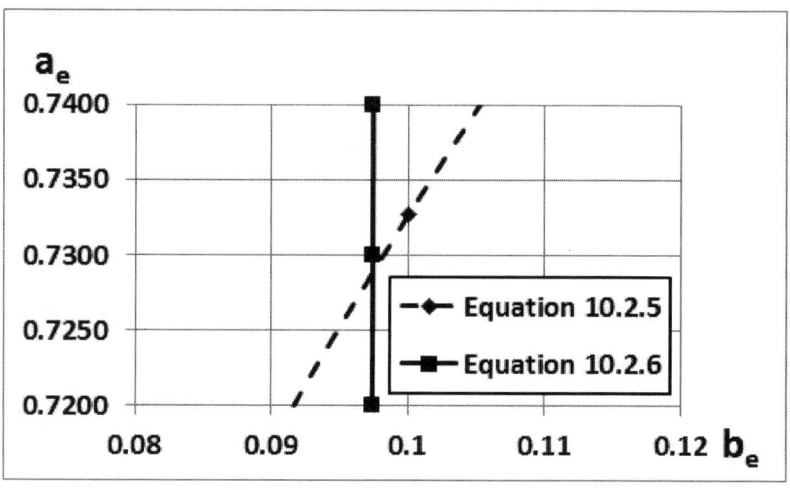

Figure 10.2.5: A close-up of Figure 10.2.4.

And, when $a_s = 1$, $b_s = 0.1$, $\nu = 0.3$, $\frac{S_{xx}}{E} = -0.15$, and $\frac{S_{yy}}{E} = -0.1$, equation (9.2.3) simplifies to

$$\frac{a_s}{a_e} = e^{\frac{\nu}{E}\frac{a_e}{a_s}\left[S_{yy}(1+\frac{2a_e}{b_e})-S_{xx}\right]} = e^{(0.3)\frac{a_e}{1}\left[(-0.1)(1+\frac{2a_e}{b_e})-(-0.15)\right]}$$

$$= e^{(0.3)a_e\left[(-0.1)+(-0.1)\frac{2a_e}{b_e}+(0.15)\right]} = e^{[(0.3)a_e(\frac{-0.2a_e}{b_e}+0.05)]}$$

Therefore, at the tip, $e^{[(0.3)a_e(\frac{-0.2a_e}{b_e}+0.05)]} - \frac{1}{a_e} = 0$.

Similarly, equation (9.2.4) simplifies to

$$\frac{b_s}{b_e} = e^{\frac{\nu}{E}\frac{b_e}{b_s}\left[S_{xx}(1+\frac{2b_e}{a_e})-S_{yy}\right]} = e^{(0.3)\frac{b_e}{0.1}\left[(-0.15)(1+\frac{2b_e}{a_e})-(-0.1)\right]}$$

$$= e^{(3)b_e\left[(-0.15)+(-0.15)\frac{2b_e}{a_e}+(0.1)\right]} = e^{[(3)b_e(-0.05-\frac{0.3b_e}{a_e})]}$$

Therefore, at the top, $e^{[(3)b_e(-0.05-\frac{0.3b_e}{a_e})]} - \frac{0.1}{b_e} = 0$.

We need to solve $e^{[(0.3)a_e(\frac{-0.2a_e}{b_e}+0.05)]} - \frac{1}{a_e} = 0$ and

$e^{[(3)b_e(-0.05-\frac{0.3b_e}{a_e})]} - \frac{0.1}{b_e} = 0$, each containing two unknowns, a_e and b_e. We employ the trial and error method described in Appendix G. However, as shown in Figure 10.2.6, we find that there is no possible solution.

In comparison, when we used Cauchy stress, in part A, we found no solution.

This is a possible limitation on the use of models based on logarithmic strain definition.

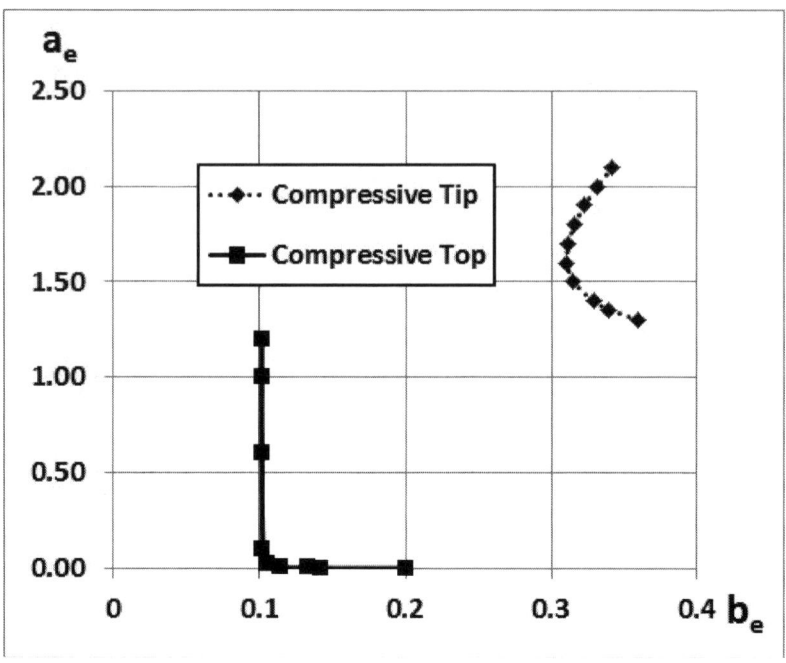

Figure 10.2.6: Plot showing two curves do not intersect.

10.3 Almansi strain

For Almansi strain, we discussed two models in Chapter 9.

Part A : With Cauchy stress,

First, we discuss the model that involved Cauchy stress.

We need to solve

$$\left[4\frac{\nu}{E}S_{yy}\left(\frac{1}{b_e}\right)\right]a_e^3 + \left[1 + 2\frac{\nu}{E}(S_{yy} - S_{xx})\right]a_e^2 - a_s^2 = 0 \quad (9.3.1)$$

and

$$\left[4\frac{\nu}{E}S_{xx}\left(\frac{1}{a_e}\right)\right]b_e^3 + \left[1 + 2\frac{\nu}{E}(S_{xx} - S_{yy})\right]b_e^2 \quad b_s^2 = 0. \quad (9.3.2)$$

When $S_{xx} = 0$, equation (9.3.2) simplifies to

$$\left[4\frac{\nu}{E}S_{xx}\left(\frac{1}{a_e}\right)\right]b_e^3 + \left[1 + 2\frac{\nu}{E}(S_{xx} - S_{yy})\right]b_e^2 - b_s^2$$

$$= \left[4(0)\left(\frac{1}{a_e}\right)\right]b_e^3 + \left[1 - 2\nu\frac{S_{yy}}{E}\right]b_e^2 - b_s^2$$

$$= \left[1 - 2\nu\frac{S_{yy}}{E}\right]b_e^2 - b_s^2 = 0.$$

Or, $b_e = b_s\sqrt{\dfrac{1}{\left[1-2\nu\frac{S_{yy}}{E}\right]}}$

With the help of this last expression we can compute b_e and then using equation (9.3.1), we can compute a_e.

For example, when $a_s = 1$, $b_s = 0.1$, $\nu = 0.3$, $\frac{S_{xx}}{E} = 0$ and $\frac{S_{yy}}{E} = 0.1$,

$$b_e = b_s\sqrt{\frac{1}{[1-2\nu\frac{S_{yy}}{E}]}} = (0.1)\sqrt{\frac{1}{[1-2(0.3)(0.1)]}} = (0.1)\sqrt{\frac{1}{[1-(0.06)]}}$$

$$= (0.1)\sqrt{\frac{1}{[0.94]}} = (0.1)1.0314212462 = 0.10314212462$$

Hence, equation (9.3.1) yields

$$\left[4\tfrac{\nu}{E}S_{yy}\left(\tfrac{1}{b_e}\right)\right]a_e^3 + \left[1 + 2\tfrac{\nu}{E}(S_{yy} - S_{xx})\right]a_e^2 - a_s^2$$

$$= \left[\tfrac{4(0.3)(0.1)}{0.10314212462}\right]a_e^3 + \left[1 + 2(0.3)(0.1) - 0\right]a_e^2 - (1)^2$$

$$= \left[\tfrac{(0.12)}{0.10314212462}\right]a_e^3 + \left[1 + (0.06)\right]a_e^2 - 1 = 0$$

Or, $[1.1634431658]a_e^3 + [1.06]a_e^2 - 1 = 0$ which can be solved by trial and error method described in section 10.1.

Table 10.3.1 is similar to the Table 10.1.1. The last row in Table 10.3.1 indicates that $a_e = 0.7248441907$.

a_e	$f(a_e)$
1	1.2234431658
0.5	-0.5895696043
0.7	-0.0815389941
0.7248441907	0.0000000000

Table 10.3.1: $f(a_e)$ values for various (a_e) values.

Thus, when $a_s = 1$, $b_s = 0.1$, $\nu = 0.3$, $\tfrac{S_{xx}}{E} = 0$, and $\tfrac{S_{yy}}{E} = 0.1$, use of Almansi strain definition and Cauchy stress yields $a_e = 0.7248441907$ and $b_e = 0.10314212462$.

In comparison, when we used logarithmic strain definition and Cauchy stress, we found $a_e = 0.66058138149$ and $b_e = 0.1030454534$.

When $\underline{S_{yy} = 0}$, equation (9.3.1) simplifies to

$$\left[4\tfrac{\nu}{E}S_{yy}\left(\tfrac{1}{b_e}\right)\right]a_e^3 + \left[1 + 2\tfrac{\nu}{E}(S_{yy} - S_{xx})\right]a_e^2 - a_s^2$$

$$= \left[4\frac{\nu}{E}(0)\left(\frac{1}{b_e}\right)\right]a_e^3 + \left[1 + 2\frac{\nu}{E}(0 - S_{xx})\right]a_e^2 - a_s^2$$

$$= \left[1 - 2\nu\frac{S_{xx}}{E}\right]a_e^2 - a_s^2 = 0$$

Or, $a_e = a_s \sqrt{\dfrac{1}{\left[1 - 2\nu\frac{S_{xx}}{E}\right]}}$

With the help of this last expression we can compute a_e and then using equation (9.3.2), we can compute b_e.

When $a_s = 1$, $b_s = 0.1$, $\nu = 0.3$, $\frac{S_{xx}}{E} = 0.1$ and $\frac{S_{yy}}{E} = 0$,

$a_e = a_s\sqrt{\dfrac{1}{[1 - 2\nu\frac{S_{xx}}{E}]}} = (1)\sqrt{\dfrac{1}{[1 - 2(0.3)(0.1)]}} = \sqrt{\dfrac{1}{[1 - (0.06)]}}$

$= \sqrt{\dfrac{1}{[0.94]}} = 1.0314212462$

Hence, equation (9.3.1) yields

$$\left[4\frac{\nu}{E}S_{xx}\left(\frac{1}{a_e}\right)\right]b_e^3 + \left[1 + 2\frac{\nu}{E}(S_{xx} - S_{yy})\right]b_e^2 - b_s^2$$

$$= \left[\left(\frac{4(0.3)(0.1)}{1.0314212462}\right)\right]b_e^3 + \left[1 + 2(0.3)(0.1 - 0)\right]b_e^2 - (0.1)^2$$

$$= \left[\frac{0.12}{1.0314212462}\right]b_e^3 + \left[1 + 0.06\right]b_e^2 - 0.01$$

$$= \left[0.11634431658\right]b_e^3 + \left[1.06\right]b_e^2 - 0.01 = 0$$

Or, $[11.634431658]b_e^3 + [106]b_e^2 - 1 = 0$ which can be solved by trial and error method described in section 10.1.

Table 10.3.2 is similar to the Table 10.1.1. The last row in Table 10.3.2 indicates that $b_e = 0.096617639855$.

---

b_e	$f(b_e)$
0.1	0.0716344317
0.05	-0.7335456960
0.09	-0.1329184993
0.096617639855	0.0000000000

Table 10.3.2: $f(b_e)$ values for various (b_e) values.

Thus, when $a_s = 1$, $b_s = 0.1$, $\nu = 0.3$, $\frac{S_{xx}}{E} = 0.1$ and $\frac{S_{yy}}{E} = 0$, use of Almansi strain definition and Cauchy stress yields $a_e = 1.0314212462$ and $b_e = 0.096617639855$.

In comparison, when we used logarithmic strain definition and Cauchy stress, we found $a_e = 1.030454534$ and $b_e = 0.0965007963$.

When both S_{xx} and S_{yy} are non-zero, we substitute values for the known quantities in equation (9.3.1) and equation (9.3.2) then solve the simplified equations.

For example, when $a_s = 1$, $b_s = 0.1$, $\nu = 0.3$, $\frac{S_{xx}}{E} = 0.15$, and $\frac{S_{yy}}{E} = 0.1$, equation (9.3.1) simplifies to

$$\left[4\frac{\nu}{E}S_{yy}\left(\frac{1}{b_e}\right)\right]a_e^3 + \left[1 + 2\frac{\nu}{E}(S_{yy} - S_{xx})\right]a_e^2 - a_s^2$$

$$= \left[\frac{4(0.3)(0.1)}{b_e}\right]a_e^3 + \left[1 + 2(0.3)(0.1 - 0.15)\right]a_e^2 - (1)^2$$

$$= \left[\frac{(0.12)}{b_e}\right]a_e^3 + \left[1 + (0.6)(-0.05)\right]a_e^2 - 1 = 0.$$

Or,

$$\left[\frac{0.12}{b_e}\right]a_e^3 + \left[0.97\right]a_e^2 - 1 = 0. \qquad (10.3.1)$$

Similarly, equation (9.3.2) simplifies to

$$\left[4\frac{\nu}{E}S_{xx}\left(\frac{1}{a_e}\right)\right]b_e^3 + \left[1 + 2\frac{\nu}{E}(S_{xx} - S_{yy})\right]b_e^2 - b_s^2$$

$$= \left[4(0.3)(0.15)\left(\frac{1}{a_e}\right)\right]b_e^3 + \left[1 + 2(0.3)(0.15 - 0.1)\right]b_e^2 - (0.1)^2$$

$$= \left[(1.2)(0.15)\left(\frac{1}{a_e}\right)\right]b_e^3 + \left[1 + (0.6)(0.05)\right]b_e^2 - (0.1)^2$$

$$= \left[\frac{0.18}{a_e}\right]b_e^3 + \left[1.03\right]b_e^2 - 0.01 = 0$$

Or,

$$\left[\frac{18}{a_e}\right]b_e^3 + \left[103\right]b_e^2 - 1 = 0. \qquad (10.3.2)$$

a_e	b_e	$Equation - 10.3.1$
0.70285333207	0.08	0.0000000000
0.71985244243	0.09	0.0000000000
0.73485261903	0.10	0.0000000000
0.74822109671	0.11	0.0000000000
0.76023441675	0.12	0.0000000000

Table 10.3.3: Five solutions to equation (10.3.1).

We solve equation (10.3.1) and equation (10.3.2) each containing two unknowns, a_e and b_e by employing the trial and error method described in Appendix G. Table 10.3.3

shows five solutions to equation (10.3.1) where b_e values are selected as 0.08, 0.09, 0.10, 0.11, and 0.12, while a_e values are found by trial and error so that equation (10.3.1) is satisfied.

Similarly, Table 10.3.4 shows five solutions to equation (10.3.2) where a_e values are selected as 0.70, 0.71, 0.72, 0.73, and 0.74, while b_e values are found by trial and error so that equation (10.3.2) is satisfied.

a_e	b_e	$Equation - 10.3.2$
0.70	0.097356879045	0.0000000000
0.71	0.097372963582	0.0000000000
0.72	0.097388613981	0.0000000000
0.73	0.097403847592	0.0000000000
0.74	0.097418680851	0.0000000000

Table 10.3.4: Five solutions to equation (10.3.2).

These ten solutions are plotted in Figure 10.3.1 which shows that the common solution is just right of the fourth point in Table 10.3.4 i.e. $a_e = 0.73$ and $b_e = 0.097403847592$.

By trial and error method, we find that $a_e = 0.73113058511$ and $b_e = 0.097405544395$ is the solution that satisfies both equation (10.3.1) and equation (10.3.2).

In comparison, when we used logarithmic strain with Cauchy stress, in part A of section 10.2, we found $a_e = 0.6709698413$ and $b_e = 0.097234704$.

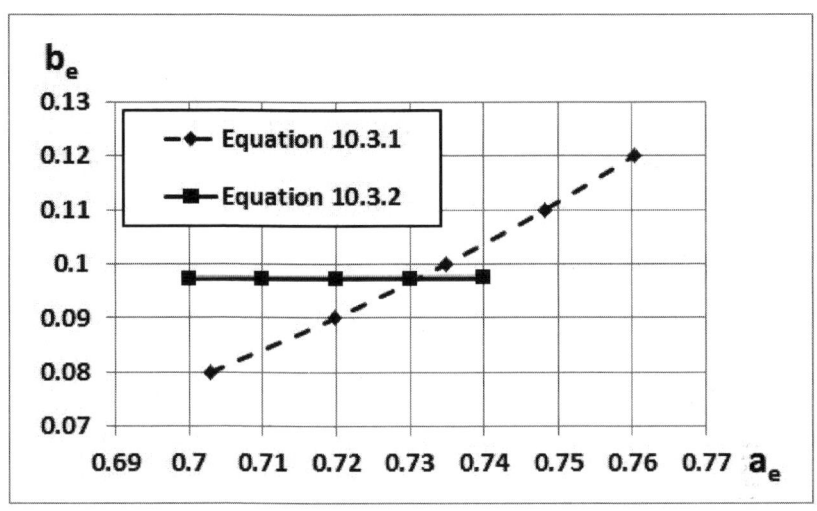

Figure 10.3.1: Plot of equation (10.3.1) and equation (10.3.2).

And, when $a_s = 1$, $b_s = 0.1$, $\nu = 0.3$, $\frac{S_{xx}}{E} = -0.15$, and $\frac{S_{yy}}{E} = -0.1$, equation (9.3.1) simplifies to

$$\left[4\frac{\nu}{E} S_{yy} \left(\frac{1}{b_e} \right) \right] a_e^3 + \left[1 + 2\frac{\nu}{E}(S_{yy} - S_{xx}) \right] a_e^2 - a_s^2$$

$$= \left[\frac{4(0.3)(-0.1)}{b_e} \right] a_e^3 + \left[1 + 2(0.3)(-0.1 + 0.15) \right] a_e^2 - (1)^2$$

$$= \left[\frac{(-0.12)}{b_e} \right] a_e^3 + \left[1 + (0.6)(0.05) \right] a_e^2 - 1 = 0.$$

Or, $\left[\frac{-0.12}{b_e} \right] a_e^3 + \left[1.03 \right] a_e^2 - 1 = 0$, at the tip.

Similarly, equation (9.3.2) simplifies to

$$\left[4\frac{\nu}{E} S_{xx} \left(\frac{1}{a_e} \right) \right] b_e^3 + \left[1 + 2\frac{\nu}{E}(S_{xx} - S_{yy}) \right] b_e^2 - b_s^2$$

$$= \left[\frac{4(0.3)(-0.15)}{a_e}\right] b_e^3 + \left[1 + 2(0.3)(-0.15 + 0.1)\right] b_e^2 - (0.1)^2$$

$$= \left[\frac{-0.18}{a_e}\right] b_e^3 + \left[1 + (0.6)(-0.05)\right] b_e^2 - (0.1)^2$$

$$= -\left[\frac{0.18}{a_e}\right] b_e^3 + \left[0.97\right] b_e^2 - 0.01 = 0, \text{ at the top.}$$

Or, $\left[\frac{18}{a_e}\right] b_e^3 = \left[97\right] b_e^2 - 1$. Or, $a_e = \left[\frac{18 b_e^3}{97 b_e^2 - 1}\right]$.

We know $a_e > 0$ and $b_e > 0$. Furthermore, for $a_e > 0$, $97 b_e^2 - 1 > 0$. Or, $b_e > \sqrt{\frac{1}{97}}$. Thus, $b_e > 0.1015346165$.

Substituting $a_e = \left[\frac{18 b_e^3}{97 b_e^2 - 1}\right]$ in $-\left[\frac{0.12}{b_e}\right] a_e^3 + (1.03) a_e^2 - 1 = 0$, we obtain $-\left[\frac{0.12}{b_e}\right]\left[\frac{18 b_e^3}{97 b_e^2 - 1}\right]^3 + \left[1.03\right]\left[\frac{18 b_e^3}{97 b_e^2 - 1}\right]^2 - 1 = 0$.

Let $f(b_e) = -[(0.12)(b_e^8)]\left[\frac{18}{97 b_e^2 - 1}\right]^3 + (1.03)\left[\frac{18 b_e^3}{97 b_e^2 - 1}\right]^2 - 1$. We need to find b_e such that $f(b_e) = 0$.

Figure 10.3.2 shows that the solution is between $b_e = 5$ and $b_e = 6$. By trial and error method, we find that $b_e = 5.3662750233981$ is the solution. Hence, $a_e = 0.99616023745$.

In comparison, when we used logarithmic strain with Cauchy stress, in part A of section 10.2, we could not find any solution.

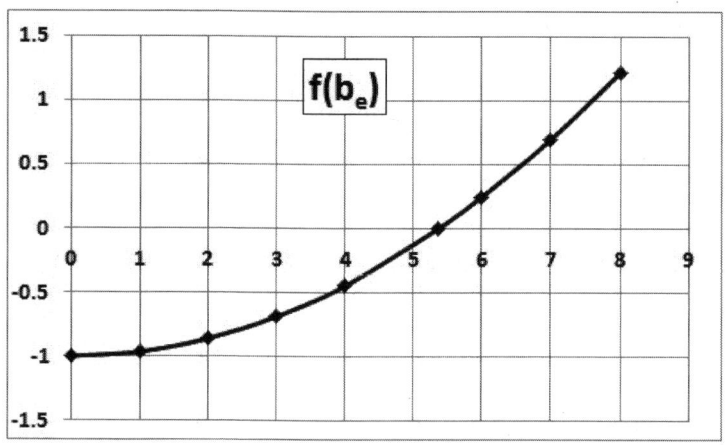

Figure 10.3.2: Plot of $f(b_e)$.

Part B : With Kirchoff stress,

Now, we discuss the model that involved Kirchoff stress. We need to solve

$$\left[4\frac{\nu}{E}S_{yy}\left(\frac{1}{b_e}\right)\right]a_e^4 + 2\frac{\nu}{E}\left[S_{yy} - S_{xx}\right]a_e^3 + a_s a_e^2 - a_s^3 = 0. \quad (9.3.3)$$

and

$$\left[4\frac{\nu}{E}S_{xx}\left(\frac{1}{a_e}\right)\right]b_e^4 + 2\frac{\nu}{E}\left[S_{xx} - S_{yy}\right]b_e^3 + b_s b_e^2 - b_s^3 = 0. \quad (9.3.4)$$

When $\underline{S_{xx} = 0}$, equation (9.3.4) simplifies to

$$\left[4\frac{\nu}{E}S_{xx}\left(\frac{1}{a_e}\right)\right]b_e^4 + 2\frac{\nu}{E}\left[S_{xx} - S_{yy}\right]b_e^3 + b_s b_e^2 - b_s^3$$

117

$$= \left[4(0)\left(\tfrac{1}{a_e}\right)\right]b_e^4 + 2\tfrac{\nu}{E}\left[0 - S_{yy}\right]b_e^3 + b_s b_e^2 - b_s^3$$

$$= -2\tfrac{\nu}{E}S_{yy}b_e^3 + b_s b_e^2 - b_s^3 = 0.$$

Or, $2\tfrac{\nu}{E}S_{yy}b_e^3 - b_s b_e^2 + b_s^3 = 0$

With the help of this last expression we can compute b_e and then using equation (9.3.3), we can compute a_e.

When $a_s = 1$, $b_s = 0.1$, $\nu = 0.3$, $\tfrac{S_{xx}}{E} = 0$ and $\tfrac{S_{yy}}{E} = 0.1$,

$2\tfrac{\nu}{E}S_{yy}b_e^3 - b_s b_e^2 + b_s^3 = 0$ simplifies to

$$2(0.3)(0.1)b_e^3 - (0.1)b_e^2 + (0.1)^3$$
$$= (0.06)b_e^3 - (0.1)b_e^2 + (0.1)^3 = 0.$$

Or, $60b_e^3 - 100b_e^2 + 1 = 0$ which can be solved by trial and error method.

b_e	$f(b_e)$
0.1	0.0600000000
0.11	-0.1301400000
0.105	-0.0330425000
0.10324924909	0.0000000000

Table 10.3.5: $f(b_e)$ values for various (b_e) values.

The last row in Table 10.3.5 shows that

$b_e = 0.10324924909$.

Hence, equation (9.3.3) simplifies to

$$\left[4\tfrac{\nu}{E}S_{yy}\left(\tfrac{1}{b_e}\right)\right]a_e^4 + 2\tfrac{\nu}{E}\left[S_{yy} - S_{xx}\right]a_e^3 + a_s a_e^2 - a_s^3$$

$$= \left[\left(\frac{4(0.3)(0.1)}{0.10324924909} \right) \right] a_e^4 + 2(0.3) \Big[(0.1) - 0 \Big] a_e^3 + (1)a_e^2 - (1)^3$$

$$= \left[\frac{(0.12)}{0.10324924909} \right] a_e^4 + (0.06)a_e^3 + a_e^2 - 1 = 0.$$

Or, $(1.1622360555)a_e^4 + (0.06)a_e^3 + a_e^2 - 1 = 0$.

The last row in Table 10.3.6 shows that

$a_e = 0.762261481$.

a_e	$f(a_e)$
1	1.2222360555
0.8	0.1467718883
0.7	-0.2103671231
0.762261481	0.0000000000

Table 10.3.6: $f(a_e)$ values for various (a_e) values.

Thus, when $a_s = 1$, $b_s = 0.1$, $\nu = 0.3$, $\frac{S_{xx}}{E} = 0$, and $\frac{S_{yy}}{E} = 0.1$, use of Almansi strain definition and Kirchoff stress yields $a_e = 0.762261481$ and $b_e = 0.10324924909$.

In comparison, when we used Almansi strain definition and Cauchy stress, we found $a_e = 0.7248441907$ and $b_e = 0.10314212462$.

When $\underline{S_{yy} = 0}$, equation (9.3.3) simplifies to

$$\left[4\frac{\nu}{E} S_{yy} \left(\frac{1}{b_e} \right) \right] a_e^4 + 2\frac{\nu}{E} \Big[S_{yy} - S_{xx} \Big] a_e^3 + a_s a_e^2 - a_s^3$$

$$= \left[4\frac{\nu}{E}(0)\left(\frac{1}{b_e} \right) \right] a_e^4 + 2\frac{\nu}{E} \Big[(0) - S_{xx} \Big] a_e^3 + a_s a_e^2 - a_s^3$$

$$= -2\frac{\nu}{E}\Big[S_{xx}\Big]a_e^3 + a_s a_e^2 - a_s^3 = 0.$$

Or, $2\nu\frac{S_{xx}}{E}a_e^3 - a_s a_e^2 + a_s^3 = 0.$

With the help of this last expression we can compute a_e and then using equation (9.3.4), we can compute b_e.

When $a_s = 1$, $b_s = 0.1$, $\nu = 0.3$, $\frac{S_{xx}}{E} = 0.1$ and $\frac{S_{yy}}{E} = 0$,

$2\nu\frac{S_{xx}}{E}a_e^3 - a_s a_e^2 + a_s^3 = 0.$ simplifies to

$2(0.3)(0.1)a_e^3 - (1)a_e^2 + (1)^3 = 0$

Or, $0.06a_e^3 - a_e^2 + 1 = 0$ which can be solved by trial and error method.

a_e	$f(a_e)$
1	0.0600000000
1.1	-0.1301400000
1.05	-0.0330425000
1.0324924909	0.0000000000

Table 10.3.7: $f(a_e)$ values for various (a_e) values.

The last row in Table 10.3.7 shows that

$a_e = 1.0324924909.$

Hence, equation (9.3.4) simplifies to

$$\Big[4\frac{\nu}{E}S_{xx}\Big(\frac{1}{a_e}\Big)\Big]b_e^4 + 2\frac{\nu}{E}\Big[S_{xx} - S_{yy}\Big]b_e^3 + b_s b_e^2 - b_s^3$$

$$= \Big[\Big(\frac{4(0.3)(0.1)}{1.0324924909}\Big)\Big]b_e^4 + 2(0.3)\Big[(0.1) - 0\Big]b_e^3 + (0.1)b_e^2 - (0.1)^3$$

$$= \left[\frac{(0.12)}{1.0324924909}\right] b_e^4 + (0.06)b_e^3 + (0.1)b_e^2 - (0.001) = 0.$$

Or, $(116.2236055541)b_e^4 + (60)b_e^3 + (100)b_e^2 - 1 = 0.$

The last row in Table 10.3.8 shows that

$b_e = 0.09672306419.$

b_e	$f(b_e)$
0.1	0.0716223606
0.09	-0.1386345692
0.095	-0.036591047
0.09672306419	0.0000000000

Table 10.3.8: $f(b_e)$ values for various (b_e) values.

Thus, when $a_s = 1$, $b_s = 0.1$, $\nu = 0.3$, $\frac{S_{xx}}{E} = 0.1$ and $\frac{S_{yy}}{E} = 0$ use of Almansi strain definition and Kirchoff stress yields $a_e = 1.0324924909$ and $b_e = 0.09672306419$.

In comparison, when we used Almansi strain definition and Cauchy stress, we found $a_e = 1.0314212462$ and $b_e = 0.096617639855$.

When both S_{xx} and S_{yy} are non-zero, we substitute values for the known quantities in equation (9.3.3) and equation (9.3.4) then solve the simplified equations.

For example, when $a_s = 1$, $b_s = 0.1$, $\nu = 0.3$, $\frac{S_{xx}}{E} = 0.15$, and $\frac{S_{yy}}{E} = 0.1$, equation (9.3.3) simplifies to

$$\left[4\frac{\nu}{E}S_{yy}\left(\frac{1}{b_e}\right)\right]a_e^4 + 2\frac{\nu}{E}\left[S_{yy} - S_{xx}\right]a_e^3 + a_s a_e^2 - a_s^3$$

$$= \left[\frac{4[0.3][0.1]}{b_e}\right]a_e^4 + 2(0.3)[(0.1) - (0.15)]a_e^3 + [1]a_e^2 - [1]^3 = 0.$$

Or,

$$\left[\frac{(0.12)}{b_e}\right]a_e^4 - (0.03)a_e^3 + a_e^2 - 1 = 0. \qquad (10.3.3)$$

Similarly, equation (9.3.4) simplifies to

$$\left[4\frac{\nu}{E}S_{xx}\left(\frac{1}{a_e}\right)\right]b_e^4 + 2\frac{\nu}{E}\left[S_{xx} - S_{yy}\right]b_e^3 + b_sb_e^2 - b_s^3$$

$$= \left[\frac{4(0.3)(0.15)}{a_e}\right]b_e^4 + 2(0.3)\left[(0.15) - (0.1)\right]b_e^3 + (0.1)b_e^2 - (0.1)^3$$

$$= 0.$$

Or,

$$\left[\frac{(0.18)}{a_e}\right]b_e^4 + (0.03)b_e^3 + (0.1)b_e^2 - 0.001 = 0. \qquad (10.3.4)$$

a_e	b_e	$Equation - 10.3.3$
0.74379932562	0.08	0.0000000000
0.7575604678	0.09	0.0000000000
0.76971698773	0.10	0.0000000000
0.78056888967	0.11	0.0000000000
0.79034033837	0.12	0.0000000000

Table 10.3.9: Five solutions to equation (10.3.3).

We solve equation (10.3.3) and equation (10.3.4) each containing two unknowns, a_e and b_e by employing the trial and error method described in Appendix G.

Table 10.3.9 shows five solutions to equation (10.3.3) where b_e values are selected as 0.08, 0.09, 0.10, 0.11, and 0.12, while a_e values are found by trial and error so that equation (10.3.3) is satisfied.

Similarly, Table 10.3.10 shows five solutions to equation (10.3.4) where a_e values are selected as 0.77, 0.78, 0.79, 0.80, and 0.81, while b_e values are found by trial and error so that equation (10.3.4) is satisfied.

a_e	b_e	$Equation - 10.3.4$
0.75	0.097494386	0.0000000000
0.76	0.09750782	0.0000000000
0.77	0.09752092	0.0000000000
0.78	0.097533696	0.0000000000
0.79	0.09754616	0.0000000000

Table 10.3.10: Five solutions to equation (10.3.4).

These ten solutions are plotted in Figure 10.3.3 which shows that the common solution is just left of the third point in Table 10.3.10 i.e. $a_e = 0.77$ and $b_e = 0.09752092$.

By trial and error method, we find that $a_e = 0.7668301778$ and $b_e = 0.0975168055$ is the solution that satisfies both equation (10.3.3) and equation (10.3.4).

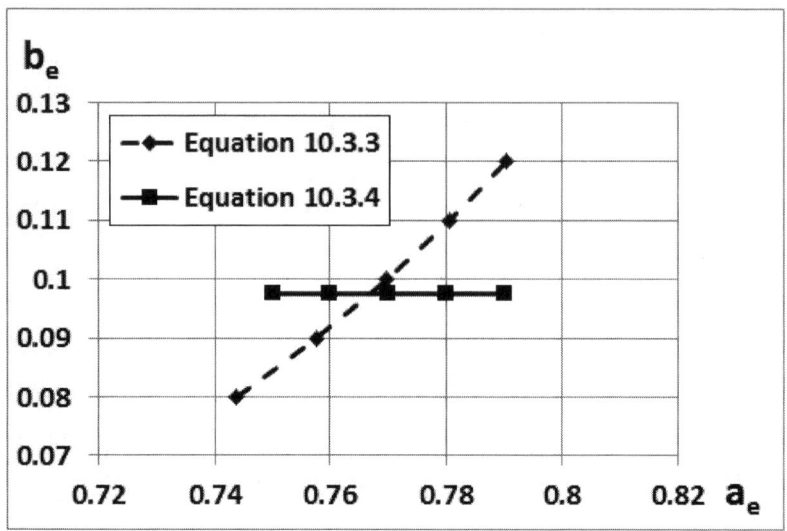

Figure 10.3.3: Plot of equation (10.3.3) and equation (10.3.4).

In comparison, when we used Almansi strain with Cauchy stress, we found $a_e = 0.73113058511$ and $b_e = 0.097405544395$.

And, when $a_s = 1$, $b_s = 0.1$, $\nu = 0.3$, $\frac{S_{xx}}{E} = -0.15$, and $\frac{S_{yy}}{E} = -0.1$, equation (9.3.3) simplifies to

$$\left[4\frac{\nu}{E}S_{yy}\left(\frac{1}{b_e}\right)\right]a_e^4 + 2\frac{\nu}{E}\left[S_{yy} - S_{xx}\right]a_e^3 + a_s a_e^2 - a_s^3$$

$$= \left[\frac{4[0.3][-0.1]}{b_e}\right]a_e^4 + 2(0.3)[(-0.1) - (-0.15)]a_e^3 + [1]a_e^2 - [1]^3$$

$$= 0.$$

Or, $\left[\frac{(-0.12)}{b_e}\right]a_e^4 + (0.03)a_e^3 + a_e^2 - 1 = 0$, at the tip.

Or, $b_e = \left[\frac{(0.12a_e^4)\cdot}{(0.03)a_e^3 + a_e^2 - 1}\right]$.

Since $b_e > 0$, $[(0.03)a_e^3 + a_e^2 - 1] > 0$ must be satisfied.

We observe that $[(0.03)a_e^3 + a_e^2 - 1] = 0$ has three roots i.e. -33.303.., -1.015.., and 0.985.. Also, $[(0.03)a_e^3 + a_e^2 - 1] > 0$ when a_e is between -33.303.. and -1.015.. or $a_e > 0.9855369$. Since $a_c > 0$, we disregard the first region. Hence, $a_e > 0.9855369$ must be satisfied.

Similarly, equation (9.3.4) simplifies to

$$\left[4\frac{\nu}{E}S_{xx}\left(\frac{1}{a_e}\right)\right]b_e^4 + 2\frac{\nu}{E}\left[S_{xx} - S_{yy}\right]b_e^3 + b_s b_e^2 - b_s^3$$

$$= \left[\frac{4(0.3)(-0.15)}{a_e}\right]b_e^4 + 2(0.3)\left[(-0.15) - (-0.1)\right]b_e^3$$

$$+ (0.1)b_e^2 - (0.1)^3 = 0.$$

Or, $\left[\frac{(-0.18)}{a_e}\right]b_e^4 - (0.03)b_e^3 + (0.1)b_e^2 - 0.001 = 0$, at the top.

Or, $a_e = \left[\frac{(0.18b_e^4)}{-(0.03)b_e^3 + (0.1)b_e^2 - 0.001}\right]$.

Since $a_e > 0$, $[-(0.03)b_e^3 + (0.1)b_e^2 - 0.001] > 0$ must be satisfied.

We observe that $[-(0.03)b_e^3 + (0.1)b_e^2 - 0.001] = 0$ has three roots i.e. -0.0985.., 0.1015.., and 3.33030..

Also, $[-(0.03)b_e^3 + (0.1)b_e^2 - 0.001] > 0$ when $b_e < -0.985..$ or b_e is between 0.1015.. and 3.3303... Since $b_e > 0$, we disregard the first region. Hence, b_e must be between 0.1015.. and 3.3303..

We attempt to solve $\left[\frac{(-0.12)}{b_e}\right]a_e^4 + (0.03)a_e^3 + a_e^2 - 1 = 0$ and

$\left[\frac{(-0.18)}{a_e}\right]b_e^4 - (0.03)b_e^3 + (0.1)b_e^2 - 0.001 = 0$ each containing two unknowns, a_e and b_e by employing the trial and error method described in Appendix G.

However, as shown in Figure 10.3.4, we find that there is no possible solution.

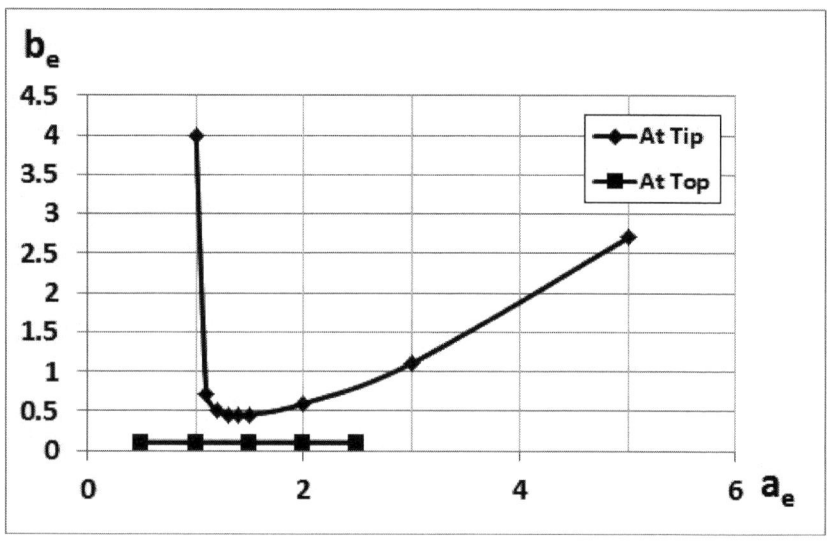

Figure 10.3.4: Plot showing two curves do not intersect.

Chapter 11

Comparing results from six models

11.1 Computing a_e and b_e for four loading conditions using Engineering strain

In order to complete a comparison for the four loading conditions we explored in Chapter 10, we return to two expressions developed for Engineering strain in Chapter 8. From equations (8.2.2) and equation (8.2.3) we know,

$$a_e = a_s b_s \left[\frac{[1 - (\frac{\nu}{E}S_{xx}) - (\frac{\nu}{E}S_{yy})][1 + (\frac{\nu}{E}S_{xx}) + (\frac{\nu}{E}S_{yy})]}{b_s(1 + \frac{\nu}{E}S_{yy} - \frac{\nu}{E}S_{xx}) + 2a_s(\frac{\nu}{E}S_{yy})} \right]$$

and

$$b_e = \frac{a_s b_s \left(1 - \frac{\nu}{E} S_{xx} - \frac{\nu}{E} S_{yy}\right)\left(1 + \frac{\nu}{E} S_{xx} + \frac{\nu}{E} S_{yy}\right)}{a_s\left[1 + \left(\frac{\nu}{E} S_{xx} - \frac{\nu}{E} S_{yy}\right)\right] + 2\left(\frac{\nu}{E} S_{xx}\right)[b_s]}.$$

When $a_s = 1$, $b_s = 0.1$, $\nu = 0.3$, $\frac{S_{xx}}{E} = 0$, and $\frac{S_{yy}}{E} = 0.1$,

$$a_e = a_s b_s \left[\frac{[1 - (\frac{\nu}{E} S_{xx}) - (\frac{\nu}{E} S_{yy})][1 + (\frac{\nu}{E} S_{xx}) + (\frac{\nu}{E} S_{yy})]}{b_s(1 + \frac{\nu}{E} S_{yy} - \frac{\nu}{E} S_{xx}) + 2a_s(\frac{\nu}{E} S_{yy})}\right]$$

$$= (1)(0.1)\left[\frac{[1 - (0) - (0.3)(0.1)][1 + (0) + (0.3)(0.1)]}{(0.1)(1 + (0.3)(0.1) - 0) + 2(1)(0.3)(0.1)}\right]$$

$$= \frac{(0.1)[1 - 0.03][1 + 0.03]}{(0.1)(1 + 0.03) + 2(0.03)} = \frac{(0.1)[0.97][1.03]}{(0.1)(1.03) + 2(0.03)}$$

$$= \frac{[0.97][0.103]}{(0.103) + (0.06)} = \frac{0.09991}{0.163} = \frac{9.991}{16.3} = 0.6129447852$$

and

$$b_e = \frac{a_s b_s \left(1 - \frac{\nu}{E} S_{xx} - \frac{\nu}{E} S_{yy}\right)\left(1 + \frac{\nu}{E} S_{xx} + \frac{\nu}{E} S_{yy}\right)}{a_s\left[1 + \left(\frac{\nu}{E} S_{xx} - \frac{\nu}{E} S_{yy}\right)\right] + 2\left(\frac{\nu}{E} S_{xx}\right)[b_s]}$$

$$= \frac{(1)(0.1)[1 - (0) - (0.3)(0.1)][1 + (0) + (0.3)(0.1)]}{(1)[1 + ((0) - (0.3)(0.1))] + (0)}$$

$$= \frac{(0.1)[1 - (0.3)(0.1)][1 + (0.3)(0.1)]}{[1 + (-(0.3)(0.1))]} = \frac{(0.1)[1 - 0.03][1 + 0.03]}{[1 - 0.03]}$$

$$= \frac{(0.1)[0.97][1.03]}{[0.97]} = (0.1)[1.03] = 0.103$$

When $a_s = 1$, $b_s = 0.1$, $\nu = 0.3$, $\frac{S_{xx}}{E} = 0.1$, and $\frac{S_{yy}}{E} = 0$,

$$a_e = a_s b_s \left[\frac{[1 - (\frac{\nu}{E} S_{xx}) - (\frac{\nu}{E} S_{yy})][1 + (\frac{\nu}{E} S_{xx}) + (\frac{\nu}{E} S_{yy})]}{b_s(1 + \frac{\nu}{E} S_{yy} - \frac{\nu}{E} S_{xx}) + 2a_s(\frac{\nu}{E} S_{yy})}\right]$$

$$= (1)(0.1)\left[\frac{[1 - ((0.3)(0.1)) - (0)][1 + ((0.3)(0.1)) + (0)]}{(0.1)(1 + 0 - (0.3)(0.1)) + 2(1)(0)}\right]$$

$$= \frac{(0.1)[0.97][1.03]}{(0.1)(0.97)} = 1.03$$

and

$$b_e = \frac{a_s b_s \left(1 - \frac{\nu}{E} S_{xx} - \frac{\nu}{E} S_{yy}\right)\left(1 + \frac{\nu}{E} S_{xx} + \frac{\nu}{E} S_{yy}\right)}{a_s\left[1 + \left(\frac{\nu}{E} S_{xx} - \frac{\nu}{E} S_{yy}\right)\right] + 2\left(\frac{\nu}{E} S_{xx}\right)[b_s]}$$

$$= \frac{(1)(0.1)(1 - (0.3)(0.1) - 0)(1 + (0.3)(0.1) + 0)}{(1)[1 + ((0.3)(0.1) - 0)] + 2((0.3)(0.1))[(0.1)]} = \frac{(0.1)(1 - 0.03)(1 + 0.03)}{[1 + 0.03] + 2(0.003)}$$

$$= \frac{(0.1)(0.97)(1.03)}{[1.036]} = \frac{(0.09991)}{[1.036]} = 0.0964382239.$$

When $a_s = 1$, $b_s = 0.1$, $\nu = 0.3$, $\frac{S_{xx}}{E} = 0.15$, and $\frac{S_{yy}}{E} = 0.1$, as shown in Appendix F, $a_e = 0.6273659306$ and $b_e = 0.0971069335$.

And when $a_s = 1$, $b_s = 0.1$, $\nu = 0.3$, $\frac{S_{xx}}{E} = -0.15$, and $\frac{S_{yy}}{E} = -0.1$, as shown in Appendix F, $a_e - 2.3960843373$ and $b_e = 0.1018826844$.

11.2 Comparing a_e and b_e for four loading conditions

These computed values for a_e and b_e using Engineering strain and computed values for a_e and b_e from Chapter 10 using Green, Logarithmic, and Almansi strain are presented in Table 11.2.1.

It is interesting to note that when $a_s = 1$, $b_s = 0.1$, $\nu = 0.3$, $\frac{S_{xx}}{E} = 0$, and $\frac{S_{yy}}{E} = 0.1$, b_e values are not too different, ranging between 0.10295 and 0.10324. On the other hand, a_e values exhibit a significant variation, ranging between 0.54843254 and 0.76226148.

However, when $a_s = 1$, $b_s = 0.1$, $\nu = 0.3$, $\frac{S_{xx}}{E} = 0.1$, and $\frac{S_{yy}}{E} = 0$, b_e values (ranging between 0.096438 and 0.09689707) as well as a_e values (ranging between 1.029563 and 1.032492) do not show much variation.

	Initial Geometry	$a_s = 1$ and $b_s = 0.1$		
	Poisson's Ratio	$\nu = 0.3$		
Load → Model ↓	$S_{xx} = 0$, $S_{yy} = 0.1E$	$S_{xx} = 0.1E$, $S_{yy} = 0$	$S_{xx} = 0.15E$, $S_{yy} = 0.1E$	$S_{xx} = -0.15E$, $S_{yy} = -0.1E$
Engineering strain Cauchy stress	$a_e = 0.6129448$ $b_e = 0.103$	$a_e = 1.03$ $b_e = 0.0964382$	$a_e = 0.6273659$ $b_e = 0.0971069$	$a_e = 2.3960843$ $b_e = 0.1018827$
Green strain Cauchy stress	$a_e = 0.5484325$ $b_e = 0.1029563$	$a_e = 1.0295630$ $b_e = 0.0968971$	$a_e = 0.5697371$ $b_e = 0.0969216$	$a_e = 1.7336895$ $b_e = 0.1021982$
Logarithmic strain Cauchy stress	$a_e = 0.6605814$ $b_e = 0.1030455$	$a_e = 1.0304545$ $b_e = 0.0965008$	$a_e = 0.6709698$ $b_e = 0.0972347$	No solution No solution
Logarithmic strain Kirchoff stress	$a_e = 0.7223679$ $b_e = 0.1031426$	$a_e = 1.0314265$ $b_e = 0.0966170$	$a_e = 0.7288431$ $b_e = 0.0974018$	No solution No solution
Almansi strain Cauchy stress	$a_e = 0.7248442$ $b_e = 0.1031421$	$a_e = 1.0314212$ $b_e = 0.0966176$	$a_e = 0.7311306$ $b_e = 0.0974055$	$a_e = 0.9961602$ $b_e = 5.3362750$
Almansi strain Kirchoff stress	$a_e = 0.7622615$ $b_e = 0.1032492$	$a_e = 1.0324925$ $b_e = 0.0967231$	$a_e = 0.7668302$ $b_e = 0.0975168$	No solution No solution

Table 11.2.1: Computed values of a_e and b_e for six models.

When $a_s = 1$, $b_s = 0.1$, $\nu = 0.3$, $\frac{S_{xx}}{E} = 0.15$, and $\frac{S_{yy}}{E} = 0.1$, $a_s = 1$ reduced to a_e ranging between 0.569737 and 0.7668302 while $b_s = 0.1$ too reduced to b_e ranging between 0.0969216 and 0.0975168.

And, when $a_s = 1$, $b_s = 0.1$, $\nu = 0.3$, $\frac{S_{xx}}{E} = -0.15$, and $\frac{S_{yy}}{E} = -0.1$, three models provided us no solutions. And out of the other three, one offered a different solution than the other two.

Thus, the model you choose can influence your results significantly under certain circumstances.

11.3 Comparing formulas for uni-axial loads

For ready reference and for an easy comparison, we enlist all six expressions for b_e when $S_{xx} = 0$.

$b_e = b_s(1 + \frac{\nu}{E}S_{yy})$ for Engineering strain with

Cauchy stress,

$b_e = b_s\sqrt{1 + 2(\frac{\nu}{E})(S_{yy})}$ for Green strain with

Cauchy stress,

$b_e = b_s e^{\frac{\nu}{E}[S_{yy}]}$ for Logarithmic strain with

Cauchy stress,

$b_e = b_s e^{\frac{\nu}{E}\frac{b_e}{b_s}[S_{yy}]}$ for Logarithmic strain

Kirchoff stress,

$b_e = b_s \sqrt{\dfrac{1}{1-2\nu \frac{S_{yy}}{E}}}$ for Almansi strain with

Cauchy stress,

$2\frac{\nu}{E} S_{yy} b_e^3 - b_s b_e^2 + b_s^3 = 0$ for Almansi strain

Kirchoff stress.

We observe that none of these expressions contain a_s. Hence, we conclude that when $S_{xx} = 0$, b_e does not depend on a_s (i.e. it does not matter whether the initial shape is a circle, ellipse, or a long narrow crack).

Similarly, when $S_{yy} = 0$

$a_e = a_s(1 + \frac{\nu}{E} S_{xx})$ for Engineering strain with

Cauchy stress,

$a_e = a_s \sqrt{1 + 2(\frac{\nu}{E})(S_{xx})}$ for Green strain with

Cauchy stress,

$a_e = a_s e^{\frac{\nu}{E}[S_{xx}]}$ for Logarithmic strain with

Cauchy stress,

$a_e = a_s e^{\frac{\nu}{E} \frac{a_e}{a_s}[S_{xx}]}$ for Logarithmic strain

Kirchoff stress,

$a_e = a_s \sqrt{\dfrac{1}{1-2\nu \frac{S_{xx}}{E}}}$ for Almansi strain with

Cauchy stress,

$2\frac{\nu}{E} S_{xx} a_e^3 - a_s a_e^2 + a_s^3 = 0$ for Almansi strain

Kirchoff stress.

Once again, we observe that none of these expressions contain b_s. Hence, we conclude that when $S_{yy} = 0$, a_e does not depend on b_s (i.e. it does not matter whether the initial shape is a circle, ellipse, or a long narrow crack).

11.4 Comparing displacements of points on the contour of the elliptical hole

Consider Green strain model when $S_{xx} = 0$. From section 10.1, we know that when $a_s = 1$, $b_s = 0.1$, $\nu = 0.3$, and $S_{yy} = 0.1E$, the initial elliptical hole deforms to another ellipse with $a_e = 0.5484325497$ and $b_e = 0.1029563014$. We can calculate c and α for each ellipse using $c = \sqrt{a^2 - b^2}$ and $\alpha = tanh^{-1}(\frac{b}{a})$. Calculated values for c_s, α_s, c_e, and α_e are shown on the top portion of Table 11.4.1.

We select ten β values between (and including) 0° and 90°. Co-ordinates of ten points corresponding to these β values, on each ellipse, are calculated using relations $x = c \, cosh\alpha \, cos\beta$ and $y = c \, sinh\alpha \, sin\beta$. Calculated values for x_s, y_s, x_e, and y_e are shown in Table 11.4.1. These values are used to plot two elliptical holes - one before deformation and the other after deformation, as shown in Figure 11.4.1.

Initial Geometry	$a_s = 1, b_s = 0.1, c_s = 0.99497437, and \alpha_s = 0.100335348$				
Load	$S_{xx} = 0 and S_{yy} = 0.1E$				
Poisson's Ratio	$\nu = 0.3$				
Final Geometry	$a_e = 0.5484325497, b_e = 0.1029563014,$ $c_e = 0.538681967, and \alpha_e = 0.189981423$				
$\beta^o s$	$\beta\,rads$	x_s	y_s	x_e	y_e
0	0	1	0	0.548432549	0
10	0.174532925	0.984807753	0.017364818	0.540100626	0.017878174
20	0.349065850	0.939692621	0.034202014	0.515358019	0.035213129
30	0.523598776	0.866025404	0.05	0.47495652	0.051478151
40	0.698131701	0.766044443	0.064278761	0.420123707	0.066179035
50	0.872664626	0.642787610	0.076604444	0.352525647	0.078869103
60	1.047197551	0.5	0.08660254	0.274216275	0.089162772
70	1.221730476	0.342020143	0.093969262	0.187574979	0.096747277
80	1.396263402	0.173648178	0.098480775	0.095234313	0.101392164
90	1.570796327	0	0.1	0	0.102956301

Table 11.4.1: Computed values of co-ordinates (x_s, y_s) and (x_e, y_e) of ten points on predeformed hole and ten points on deformed hole for ten β values for Green strain model.

Figure 11.4.2 shows displacements in the vicinity of the tip, pointing towards the centre. They have very large horizontal components due to a significant decrease in a. At the tip, the displacement is strictly horizontal. As you move from the tip to the top, vertical components of displacements increase and horizontal components decrease.

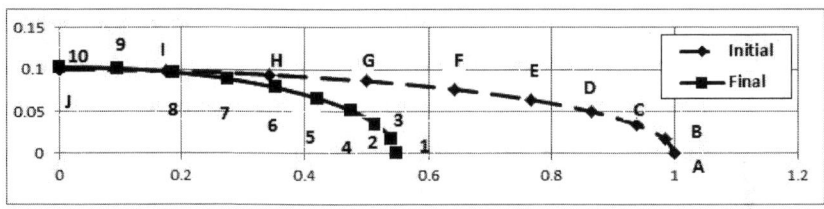

Figure 11.4.1: Graphs showing initial and final contours.

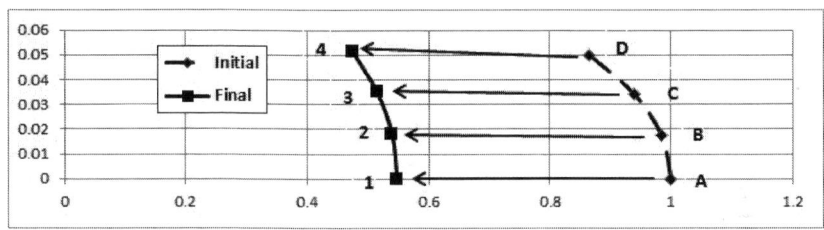

Figure 11.4.2: Displacements in the vicinity of the tip.

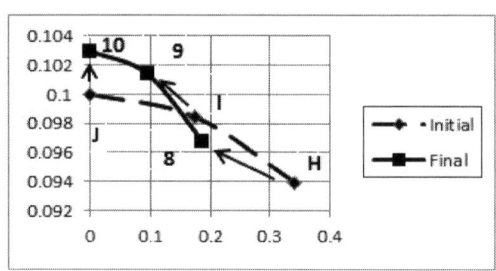

Figure 11.4.3: Displacements in the vicinity of the top.

Figure 11.4.3 shows displacements in the vicinity of the top, pointing away from the centre. They have increasing vertical components due to a relatively small increase in b. As you move closer to the top, horizontal components decrease. At the top, the displacement is strictly vertical. This is an example of contraction along major axis and expansion along minor axis (which was also discussed in Chapter 2) where a uniaxial load perpendicular to the major axis can lead to a significant decrease in focal length c which in turn causes ellipses and hyperbolas to shift towards the center thereby pulling everything towards the center. On the other hand, there is an increase in α which does not affect hyperbolas, yet pushes everything away from the center. At the tip, effect due to the decrease in c is very prominent which explains why displacements in the vicinity of the tip have large horizontal components.

It should be noted (from column 2 under $S_{xx} = 0$ in Table 11.2.1) that all six models show contraction along major axis and expansion along minor axis.

We now consider Green strain model when $S_{yy} = 0$. From section 10.1, we find that when $a_s = 1$, $b_s = 0.1$, $\nu = 0.3$, and $S_{xx} = 0.1E$, the initial elliptical hole deforms to another elliptical hole with $a_e = 1.02956301409$ and $b_e = 0.0968970714$. We carry out calculations (not shown) similar to those shown in Table 11.4.1 and plot them to obtain Figure 11.4.4.

Figure 11.4.5 shows displacements in the vicinity of the tip, pointing away the centre. They have large horizontal components due to an increase in a. At the tip, the displacement is strictly horizontal. As you move from the tip to the top, vertical components of displacements increase and horizontal components decrease.

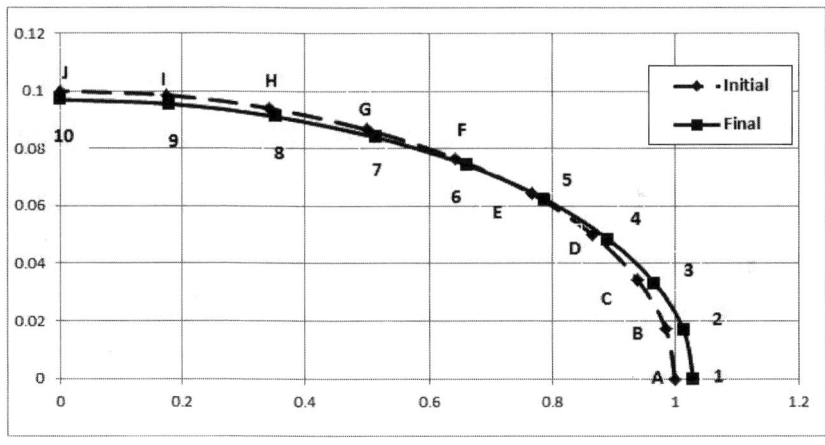

Figure 11.4.4: Graphs showing initial and final contours.

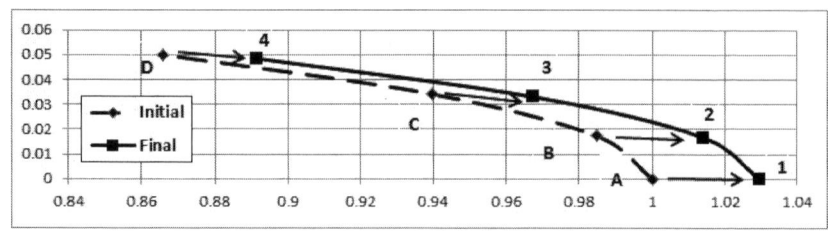

Figure 11.4.5: Displacements in the vicinity of the tip.

Figure 11.4.6 shows displacements in the vicinity of the top, pointing towards the centre. They have increasing vertical components due to a relatively small decrease in b. As you move closer to the top, horizontal components decrease. At the top, the displacement is strictly vertical.

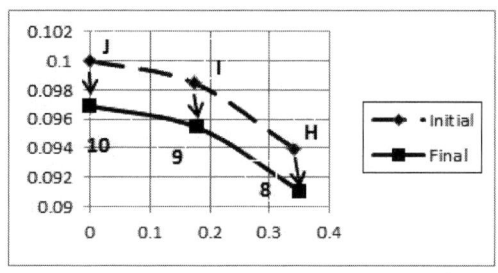

Figure 11.4.6: Displacements in the vicinity of the top.

This is an example of expansion along major axis and contraction along minor axis where a uniaxial load parallel to the major axis can lead to an increase in focal length c (from 0.994987437 to 1.02499315) which in turn causes ellipses and hyperbolas to shift away from the center thereby pushing everything away from the center. On the other hand, there is a decrease in α (from 0.100335348 to 0.094394118) which does not affect hyperbolas, yet pulls everything towards the center. At the tip, effect due to the increase in c is very prominent which explains why displacements in the vicinity of the tip have large horizontal components.

It should be noted (from column 3 under $S_{yy} = 0$ in Table 11.2.1) that all six models show expansion along major axis and contraction along minor axis.

Next, we consider Green strain model when $S_{xx} = 0.15E$ and $S_{yy} = 0.1E$. From section 10.1, we find that when $a_s = 1$, $b_s = 0.1$, and $\nu = 0.3$, the initial elliptical hole deforms to another elliptical hole with $a_e = 0.5697370473$ and $b_e = 0.0969215694537$. We carry out calculations (not shown) similar to those shown in Table 11.4.1 and plot them to obtain Figure 11.4.7.

Figure 11.4.8 shows displacements in the vicinity of the tip, pointing towards the centre. They have large horizontal components due to a decrease in a. At the tip, the displacement is strictly horizontal. As you move from the tip to the top, vertical components of displacements increase and horizontal components decrease.

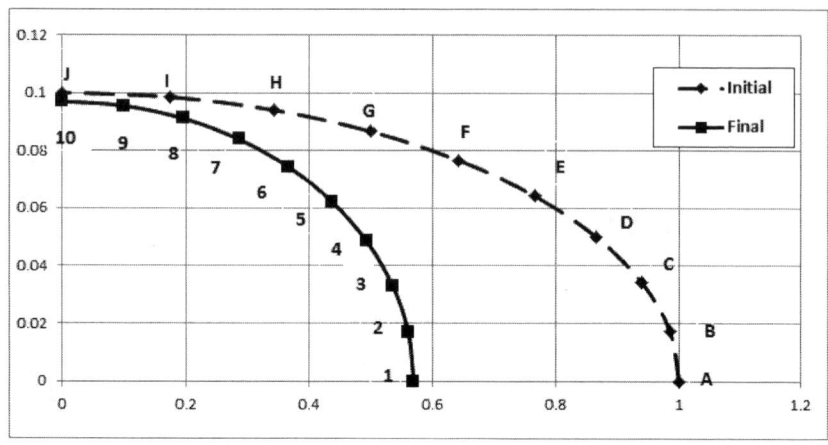

Figure 11.4.7: Graphs showing initial and final contours.

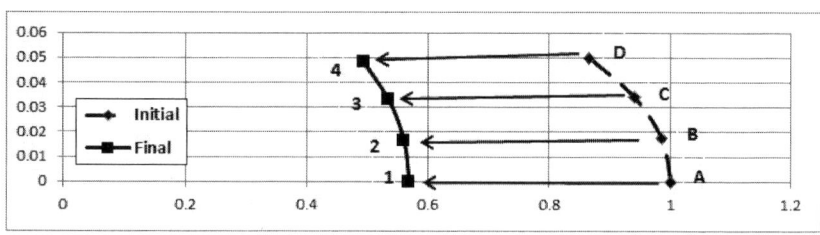

Figure 11.4.8: Displacements in the vicinity of the tip.

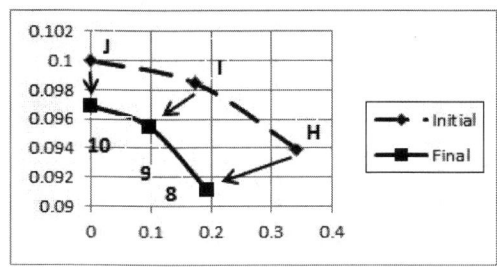

Figure 11.4.9: Displacements in the vicinity of the top.

Figure 11.4.9 shows displacements in the vicinity of the top, pointing towards the centre. They have increasing vertical components due to a relatively small increase in b. As you move closer to the top, horizontal components decrease. At the top, the displacement is strictly vertical.

This is an example of contraction along major axis and contraction along minor axis where a biaxial load can lead to a decrease in focal length c (from 0.994987437 to 0.561432554) which in turn causes ellipses and hyperbolas to shift towards the center thereby pulling everything towards the center. On the other hand, there is an increase in α (from 0.100335348 to 0.171786447) which does not affect hyperbolas, yet pushes everything away from the center. At the tip, effect due to the decrease in c is very prominent which explains why displacements in the vicinity of the tip have large horizontal components.

It should be noted (from column 3 under $S_{xx} = 0.15E$ and $S_{yy} = 0.1E$ in Table 11.2.1) that all six models show contraction along major axis and contraction along minor axis.

Next, we consider Green strain model when $S_{xx} = -0.15E$ and $S_{yy} = -0.1E$. From section 10.1, we find that when $a_s = 1$, $b_s = 0.1$, and $\nu = 0.3$, the initial elliptical hole deforms to another elliptical hole with $a_e = 1.7336895082$ and $b_e = 0.10219819$. We carry out calculations (not shown) similar to those shown in Table 11.4.1 and plot them to obtain Figure 11.4.10.

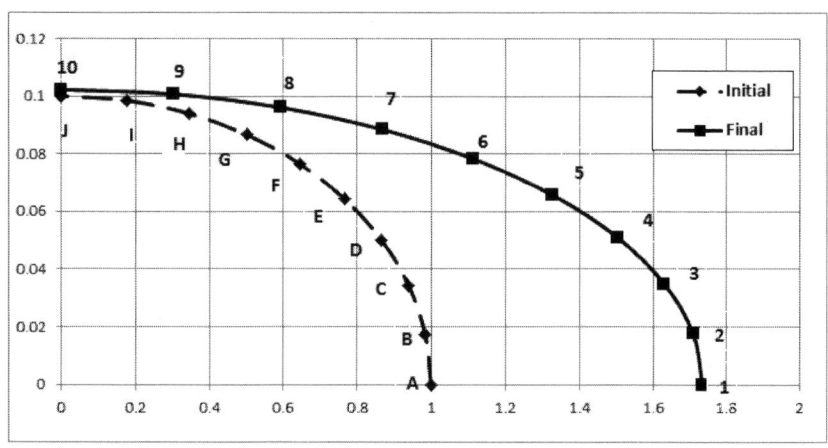

Figure 11.4.10: Graphs showing initial and final contours.

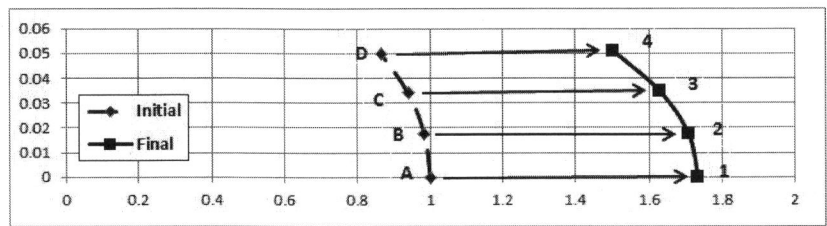

Figure 11.4.11: Displacements in the vicinity of the tip.

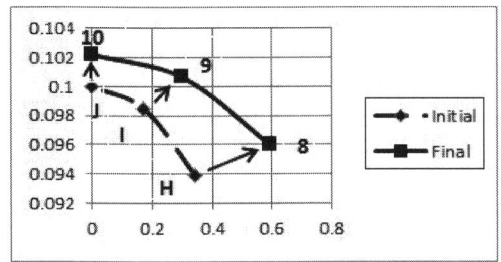

Figure 11.4.12: Displacements in the vicinity of the top.

Figure 11.4.11 shows displacements in the vicinity of the tip, pointing away from the centre. They have large horizontal components due to an increase in a. At the tip, the displacement is strictly horizontal. As you move from the tip to the top, vertical components of displacements increase and horizontal components decrease.

Figure 11.4.12 shows displacements in the vicinity of the top, pointing away from the centre. They have increasing vertical components due to a relatively small increase in b.

As you move closer to the top, horizontal components decrease. At the top, the displacement is strictly vertical.

This is an example of expansion along major axis and expansion along minor axis where a biaxial load can lead to an increase in focal length c (from 0.994987437 to 1.730674678) which in turn causes ellipses and hyperbolas to shift away from the center thereby pushing everything away from the center. On the other hand, there is a decrease in α (from 0.100335348 to 0.059016804) which does not affect hyperbolas, yet pulls everything towards the center. At the tip, effect due to the increase in c is very prominent which explains why displacements in the vicinity of the tip have large horizontal components.

11.5   Comparing zero stress points

From Appendix H and Chapter 3 of the first book, we know that the co-ordinates of the zero stress point are

$$
\left[\tanh^{-1}\left(\frac{b_e}{a_e}\right), \frac{1}{2}\cos^{-1}\left(\frac{a_e - b_e}{a_e + b_e} - \frac{(S_{yy} + S_{xx})}{(S_{yy} - S_{xx})}\frac{2a_e b_e}{(a_e + b_e)^2}\right)\right].
$$
$$(11.5.1)$$

Hence, when $S_{xx} = 0$ and $S_{yy} = 0.1E$, this expression simplifies to $\left[\tanh^{-1}\left(\frac{b_e}{a_e}\right), \frac{1}{2}\cos^{-1}\left(\frac{a_e^2 - b_e^2 - 2a_e b_e}{(a_e + b_e)^2}\right)\right]$

$$= \left[\tanh^{-1}\left(\frac{b_e}{a_e}\right), \frac{1}{2}\cos^{-1}\left(\frac{2a_e^2}{(a_e + b_e)^2} - 1\right)\right]$$

$$= \left[\tanh^{-1}\left(\frac{b_e}{a_e}\right), \frac{1}{2}\cos^{-1}\left(\frac{2}{(1 + \frac{b_e}{a_e})^2} - 1\right)\right]. \qquad (11.5.2)$$

Consequently, when $S_{xx} = 0$ and $b_e = 0$, the zero stress point is at $\alpha_e = 0$ and $\beta = 0$. However, b_e and α_e must be greater than 0.

Thus, when $S_{xx} = 0$ and b_e is very small, the zero stress point is very close to the tip. Also, there is high stress concentration near the tip. As b_e increases ($\frac{b_e}{a_e}$ increases), the zero stress point moves away from the tip and stress concentration near the tip dissipates.

Equation (11.5.2) also reveals that $\frac{b_e}{a_e}$ is the only factor that affects the zero stress point when $S_{xx} = 0$ or $S_{yy} = 0$.

The second column in Table 11.5.1 shows the values for a_e and b_e which were computed in Chapter 10 and presented in Table 11.2.1 when $S_{xx} = 0$ and $S_{yy} = 0.1E$. The third column shows the values for b_e/a_e whereas the fourth column shows the values for $c_e = \sqrt{a_e^2 - b_e^2}$ and $\alpha_e = \tanh^{-1}(\frac{b_e}{a_e})$.

Now, using Equation (11.5.2), we can calculate
$$\beta_{zsp} = \frac{1}{2}\cos^{-1}\left(\frac{2}{(1 + \frac{b_e}{a_e})^2} - 1\right).$$

We note that when $S_{xx} = 0$ and $S_{yy} = 0.1E$, at the tip, tensile $\sigma_{yy} = [S_{yy}]\left[1 + \frac{2a_e}{b_e}\right] - [S_{xx}] = [0.1E]\left[1 + \frac{2a_e}{b_e}\right]$ and, at the top, compressive $\sigma_{xx} = [S_{xx}]\left[1 + \frac{2b_e}{a_e}\right] - [S_{yy}] = -0.1E$.

Initial Geometry	$a_s = 1$, $b_s = 0.1$, $c_s = 0.994987437$, and $\alpha_s = 0.100335348$		
Load	$S_{xx} = 0$ and $S_{yy} = 0.1E$		
Poisson's Ratio	$\nu = 0.3$		
Model			
Engineering strain Cauchy stress	$a_e = 0.6129447852$ $b_e = 0.103$	$b_e/a_e = 0.168041237$	$c_e = 0.60422869$ $\alpha_e = 0.169650296$
Green strain Cauchy stress	$a_e = 0.5484325497$ $b_e = 0.1029563014$	$b_e/a_e = 0.187728284$	$c_e = 0.538681967$ $\alpha_e = 0.189981423$
Logarithmic strain Cauchy stress	$a_e = 0.6605813814$ $b_e = 0.1030454534$	$b_e/a_e = 0.155992064$	$c_e = 0.652494748$ $\alpha_e = 0.157276144$
Logarithmic strain Kirchoff stress	$a_e = 0.7223678784$ $b_e = 0.1031426499$	$b_e/a_e = 0.142784103$	$c_e = 0.714966395$ $\alpha_e = 0.143766476$
Almansi strain Cauchy stress	$a_e = 0.7248441907$ $b_e = 0.1031421246$	$b_e/a_e = 0.14229558$	$c_e = 0.717468329$ $\alpha_e = 0.143267821$
Almansi strain Kirchoff stress	$a_e = 0.762261481$ $b_e = 0.1032492490$	$b_e/a_e = 0.135451222$	$c_e = 0.755236491$ $\alpha_e = 0.136288838$

Table 11.5.1: Computed values of c_e and α_e for six models.

From Table 11.5.1, we find that the variation in $\frac{b_e}{a_e}$ values is much more than the variation in b_e values because of larger variation in a_e values. We note that $\frac{b_e}{a_e}$ value is the lowest for Almansi strain with Kirchoff stress and the highest for Green strain with Cauchy stress.

As a result, as shown in the second column in Table 11.5.2, β_{zsp} is the lowest for Almansi strain with Kirchoff stress and the highest for Green strain with Cauchy stress. The third column in Table 11.5.2 shows computed values for co-ordinates for initial points x_s and y_s using $x_s = c_s \, sinh(\alpha_s) \, sin(\beta_{zsp})$ and $y_s = c_s \, cosh(\alpha_s) \, cos(\beta_{zsp})$. The fourth column in Table 11.5.2 shows computed values for co-ordinates for zero stress points x_{zsp} and y_{zsp} using $x_{zsp} = c_e \, sinh(\alpha_e) \, sin(\beta_{zsp})$ and $y_{zsp} = c_e \, cosh(\alpha_e) \, cos(\beta_{zsp})$. And, the last column in Table 11.5.2 shows computed values for the magnitudes of vectors from (x_s, y_s) to (x_{zsp}, y_{zsp}).

We note that the magnitude of this vector is the smallest for Almansi strain with Kirchoff stress and the largest for Green strain with Cauchy stress. We also observe that the magnitude of this vector increases as β_{zsp} (as well as $\frac{b_e}{a_e}$) increases.

Table 11.5.3 shows the values for β_{zsp} and magnitudes of corresponding displacements which were calculated using a_e and b_e computed in Chapter 10 and presented in Table 11.2.1 when $S_{xx} = 0.1E$ and $S_{yy} = 0$. Also, at the tip, compressive $\sigma_{yy} = [S_{yy}] \left[1 + \frac{2a_e}{b_e}\right] - [S_{xx}] = -[0.1E]$ and, at the top, tensile $\sigma_{xx} = [S_{xx}] \left[1 + \frac{2b_e}{a_e}\right] - [S_{yy}] = [0.1E] \left[1 + \frac{2b_e}{a_e}\right]$.

Initial Geometry	$a_s = 1$, $b_s = 0.1$, $c_s = 0.99497437$, and $\alpha_s = 0.100335348$			
Load	$S_{xx} = 0$ and $S_{yy} = 0.1E$			
Poisson's Ratio	$\nu = 0.3$			
Model	β_{zsp} in degrees	Initial Points	ZSPs	Magnitudes
Engineering strain Cauchy stress	31.114742	$x_s = 0.856134$ $y_s = 0.051675$	$x_{zsp} = 0.524763$ $y_{zsp} = 0.053226$	0.331375
Green strain Cauchy stress	32.654089	$x_s = 0.841943$ $y_s = 0.053957$	$x_{zsp} = 0.461750$ $y_{zsp} = 0.055552$	0.380198
Logarithmic strain Cauchy stress	30.110689	$x_s = 0.865058$ $y_s = 0.050167$	$x_{zsp} = 0.571441$ $y_{zsp} = 0.051695$	0.293621
Logarithmic strain Kirchoff stress	28.948405	$x_s = 0.875056$ $y_s = 0.048402$	$x_{zsp} = 0.632112$ $y_{zsp} = 0.049923$	0.242948
Almansi strain Cauchy stress	28.904074	$x_s = 0.875431$ $y_s = 0.048334$	$x_{zsp} = 0.634551$ $y_{zsp} = 0.049853$	0.240884
Almansi strain Kirchoff stress	28.272217	$x_s = 0.880707$ $y_s = 0.047366$	$x_{zsp} = 0.671329$ $y_{zsp} = 0.048905$	0.209384

Table 11.5.2: Computed values of β_{zsp}s, co-ordinates (x_{zs}, y_{zs}) of zero stress points (ZSPs), and magnitudes of corresponding displacements for six models when $S_{xx} = 0$ and $S_{yy} = 0.1E$.

However, variation in β_{zsp} values in Table 11.5.3 is not as much as in Table 11.5.2 where $S_{xx} = 0$ and $S_{yy} = 0.1E$. Consequently, there is not much variation in magnitudes of corresponding displacements too.

Initial Geometry	$a_s = 1$ and $b_s = 0.1$	
Load	$S_{xx} = 0.1E$ and $S_{yy} = 0$	
Poisson's Ratio	$\nu = 0.3$	
Model	β_{zsp} in degrees	Magnitudes of displacements
Engineering strain Cauchy stress	23.881236	0.027469
Green strain Cauchy stress	23.938585	0.027049
Logarithmic strain Cauchy stress	23.883535	0.027883
Logarithmic strain Kirchoff stress	23.886424	0.028767
Almansi strain Cauchy stress	23.886553	0.028763
Almansi strain Kirchoff stress	23.887134	0.029739

Table 11.5.3: Computed values of β_{zsp}s and magnitudes of corresponding displacements for six models when $S_{xx} = 0.1E$ and $S_{yy} = 0$.

Initial Geometry	$a_s = 1$ and $b_s = 0.1$	
Load	$S_{xx} = 0.15E$ and $S_{yy} = 0.1E$	
Poisson's Ratio	$\nu = 0.3$	
Model	$a_e =$ $b_e =$	σ_{yy}/E at the tip σ_{xx}/E at the top
Engineering strain Cauchy stress	$a_e = 0.627366$ $b_e = 0.097107$	$\sigma_{yy}/E = 1.2421$ $\sigma_{xx}/E = 0.0964$
Green strain Cauchy stress	$a_e = 0.569737$ $b_e = 0.096922$	$\sigma_{yy}/E = 1.1257$ $\sigma_{xx}/E = 0.1010$
Logarithmic strain Cauchy stress	$a_e = 0.670970$ $b_e = 0.097235$	$\sigma_{yy}/E = 1.3301$ $\sigma_{xx}/E = 0.0935$
Logarithmic strain Kirchoff stress	$a_e = 0.728843$ $b_e = 0.097402$	$\sigma_{yy}/E = 1.4466$ $\sigma_{xx}/E = 0.0901$
Almansi strain Cauchy stress	$a_e = 0.731131$ $b_e = 0.097406$	$\sigma_{yy}/E = 1.4512$ $\sigma_{xx}/E = 0.0900$
Almansi strain Kirchoff stress	$a_e = 0.766830$ $b_e = 0.097517$	$\sigma_{yy}/E = 1.5223$ $\sigma_{xx}/E = 0.0882$

Table 11.5.4: Computed values of a_e, b_e, σ_{yy}/E *at the tip* and σ_{xx}/E *at the top* for six models when $S_{xx} = 0.15E$ and $S_{yy} = 0.1E$.

When $S_{xx} = 0.15E$ and $S_{yy} = 0.1E$,

$$\frac{1}{2}\cos^{-1}\left(\frac{a_e - b_e}{a_e + b_e} - \frac{(S_{yy} + S_{xx})}{(S_{yy} - S_{xx})}\frac{2a_e b_e}{(a_e + b_e)^2}\right)$$

$$= \frac{1}{2}\cos^{-1}\left(\frac{a_e - b_e}{a_e + b_e} - \frac{(0.1E + 0.15E)}{(0.1E - 0.15E)}\frac{2a_e b_e}{(a_e + b_e)^2}\right)$$

$$= \frac{1}{2}\cos^{-1}\left(\frac{a_e^2 - b_e^2 + 10a_eb_e}{(a_e + b_e)^2}\right)$$

$$= \frac{1}{2}\cos^{-1}\left(\frac{a_e^2 + 2a_eb_e + b_e^2 - 2b_e^2 + 8a_eb_e}{(a_e + b_e)^2}\right)$$

$$= \frac{1}{2}\cos^{-1}\left(1 + \frac{8a_eb_e - 2b_e^2}{(a_e + b_e)^2}\right) = \frac{1}{2}\cos^{-1}\left(1 + 2\frac{b_e}{a_e}\frac{4 - \frac{b_e}{a_e}}{(1 + \frac{b_e}{a_e})^2}\right)$$

We know $\frac{b_e}{a_e} > 0$ and cannot be greater than 1. Hence, as $\frac{b_e}{a_e}$ approaches 0, β_{zsp} approaches $\frac{1}{2}\cos^{-1}(1+)$. And, as $\frac{b_e}{a_e}$ approaches 1, β_{zsp} approaches $\frac{1}{2}\cos^{-1}(2.5)$. Consequently, there cannot be a zero stress point.

The second column in Table 11.5.4 shows the values for a_e and b_e which were computed in Chapter 10 and presented in Table 11.2.1 whereas the last column shows σ_{yy} at the tip and σ_{xx} at the top when $S_{xx} = 0.15E$ and $S_{yy} = 0.1E$.

The last column in Table 11.5.4 indicates that there is tensile stress at the tip as well as at the top for all models. Hence, there is no zero stress point.

When $S_{xx} = -0.15E$ and $S_{yy} = -0.1E$,

$$\frac{1}{2}\cos^{-1}\left(\frac{a_e - b_e}{a_e + b_e} - \frac{(S_{yy} + S_{xx})}{(S_{yy} - S_{xx})}\frac{2a_eb_e}{(a_e + b_e)^2}\right)$$

$$= \frac{1}{2}\cos^{-1}\left(\frac{a_e - b_e}{a_e + b_e} - \frac{(-0.1E - 0.15E)}{(-0.1E + 0.15E)}\frac{2a_eb_e}{(a_e + b_e)^2}\right)$$

$$= \frac{1}{2}\cos^{-1}\left(\frac{a_e^2 - b_e^2 + 10a_eb_e}{(a_e + b_e)^2}\right)$$

which is same as when $S_{xx} = 0.15E$ and $S_{yy} = 0.1E$. Hence, there cannot be a zero stress point.

The second column in Table 11.5.5 shows the values for a_e and b_e which were computed in Chapter 10 and presented in Table 11.2.1 whereas the last column shows σ_{yy} at the tip and σ_{xx} at the top when $S_{xx} = -0.15E$ and $S_{yy} = -0.1E$.

Initial Geometry	$a_s = 1 \ and \ b_s = 0.1$	
Load	$S_{xx} = -0.15E \ and \ S_{yy} = -0.1E$	
Poisson's Ratio	$\nu = 0.3$	
Model	$a_e =$ $b_e =$	σ_{yy}/E *at the tip* σ_{xx}/E *at the top*
Engineering strain *Cauchy stress*	$a_e = 2.396084$ $b_e = 0.101883$	$\sigma_{yy}/E = -4.6536$ $\sigma_{xx}/E = -0.0628$
Green strain *Cauchy stress*	$a_e = 1.733690$ $b_e = 0.102198$	$\sigma_{yy}/E = -3.3428$ $\sigma_{xx}/E = -0.0677$
Logarithmic strain *Cauchy stress*	*No solution* *No solution*	*No solution* *No solution*
Logarithmic strain *Kirchoff stress*	*No solution* *No solution*	*No solution* *No solution*
Almansi strain *Cauchy stress*	$a_e = 0.996160$ $b_e = 5.366275$	$\sigma_{yy}/E = 0.0129$ $\sigma_{xx}/E = -1.6661$
Almansi strain *Kirchoff stress*	*No solution* *No solution*	*No solution* *No solution*

Table 11.5.5: Computed values of a_e, b_e, σ_{yy}/E *at the tip* and σ_{xx}/E *at the top* for six models when $S_{xx} = -0.15E$ and $S_{yy} = -0.1E$.

As discussed earlier in Chapter 10, Logarithmic strain with Cauchy stress, Logarithmic strain with Kirchoff stress, and Almansi strain with Kirchoff stress did not yield any solutions for a_e and b_e. Consequently, we cannot use these three models to find the zero stress points. Almansi strain with Cauchy stress yielded a very different solution for a_e and b_e from the other two models. Due to an increase from $b_s = 0.1$ to $b_e = 5.366275$, this solution is far off the other two solutions. Additionally, although $S_{xx} = -0.15E$ and $S_{yy} = -0.1E$, it shows tensile σ_{yy} at the tip and hence it is not an acceptable solution.

The last column in Table 11.5.5 indicates that there is compressive stress at the tip as well as at the top for the other two models. Hence, there is no zero stress point.

11.6 Comparing responses to the far field stress at the tip

For comparison purpose, we select the first loading condition where $S_{xx} = 0$. However, instead of keeping S_{yy} at a fixed level, we vary it from 0 to $0.5E$. Since $S_{xx} = 0$, we can also include Singh, Glinka, and Dubey model in our comparison.

In order to compute σ_{yy} at the tip, we need a_e and b_e. Since $S_{xx} = 0$, we can use simplified formulas described in Section 11.3. Hence, we calculate b_e first and then calculate a_e using methods described in Chapter 10. For Singh et al model, their equation 33 ($a_f = a_i \frac{1}{(S/E)+1}$) and equation 34 ($b_f = [b_i + a_i \frac{2S/E}{1+(S/E)}]\frac{1}{1-(S/E)}$) mentioned on page 485 of their paper can be used.

Finally, we calculate tip stress σ_{yy}/E at various S_{yy}/E values. These tip stress values are plotted in Figure 11.6.1. This plot in colour is also shown on the front cover of this book.

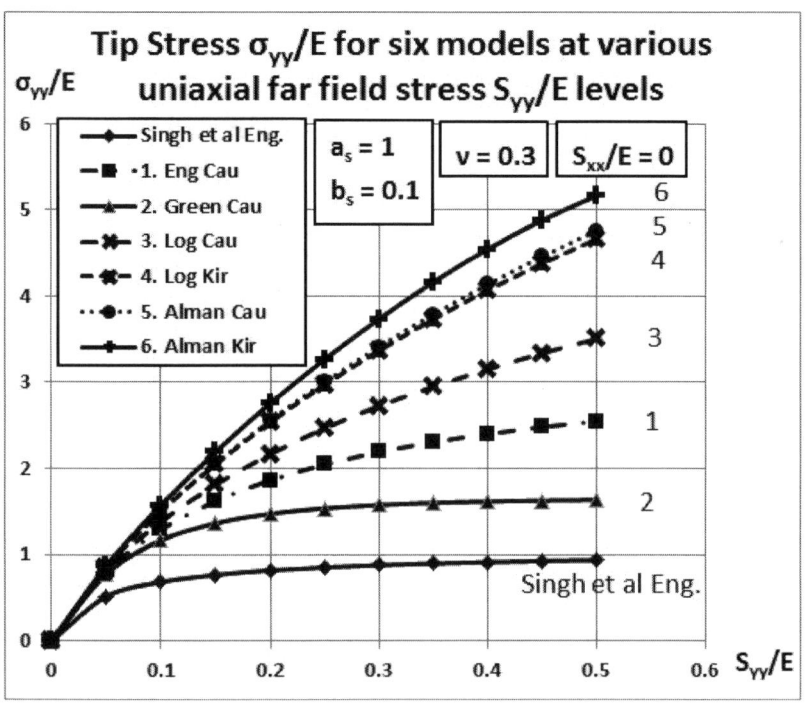

Figure 11.6.1: Tip stress.

From Figure 11.6.1, we observe that the Singh et al model yields the lowest tip stress. In section 8.4, it was shown that the Singh et al model is a special case of the general solution we developed using engineering strain definition with Cauchy stress. Furthermore, this special case can be arrived at by letting $\nu = 1$ in the general solution (Model 1 in Figure 11.6.1). Thus, the Singh, Glinka, and Dubey model can be used only for materials with $\nu = 1$. Hence, we will not discuss the Singh, Glinka, and Dubey model any further.

Among the other six models, Model 2 with Green strain and Cauchy stress definitions shows the lowest tip stress (1.64E when the far field stress is 0.5E) whereas Models 6 with Almansi strain and Kirchoff stress definitions shows the highest tip stress (5.17E when the far field stress is 0.5E).

Furthermore, Model 4 with Logarithmic strain and Kirchoff stress definitions shows tip stress values just below corresponding tip stress values for Model 5 Almansi strain and Cauchy stress definitions.

Also, at every far field stress (S_{yy}/E) value, the difference between tip stress σ_{yy}/E values for Models 5 and 6 is smaller than the corresponding difference between tip stress σ_{yy}/E values for Models 3 and 4.

It should be noted that the differences in tip stress values shown above are strictly due to the definitions of strain and stress we employ. The stress-strain relations have no influence here since, in this book, we are considering only linear stress-strain relations where E is constant.

11.7 Limitations of models

The first limitation is the self imposed limitation that E is constant. Since in this book, the focus was on understanding non-linearity in strain-displacement relations exclusively, non-linearity in stress-strain relations was not considered. Consequently, all models developed here are applicable to only linear portions of stress-strain relations irrespective of overall material behaviour. In this book, we considered far-field stresses and induced stresses as some fraction (or multiple) of E.

The second limitation is due to the nature of mathematical relations we employed in defining strain. The incompatibility of these relations with loading conditions and/or initial geometry can be a limitation.

For example, we know (from column 4 in Table 11.2.1 under $S_{xx} = -0.15E$ and $S_{yy} = -0.1E$) that not all six models showed expansion along major axis and expansion along minor axis. Three of the six models considered here provided no solutions. They are 1) model with logarithmic strain and Cauchy stress, 2) model with logarithmic strain and Kirchoff stress, and 3) model with Almansi strain and Kirchoff stress. Furthermore, when we compared zero stress points, we found that the model with Almansi strain and Cauchy stress yielded a result that was inconsistent with loading conditions. Only 1) model with engineering strain and Cauchy stress and 2) model with Green strain and Cauchy stress provided solutions that were consistent with loading conditions.

We know that the model with logarithmic strain and Cauchy stress provided us a solution when $S_{xx} = 0.15E$ and $S_{yy} = 0.1E$ but failed to provide a solution when $S_{xx} = -0.15E$ and $S_{yy} = -0.1E$. Let us explore the difference.

In Chapter 10, when $S_{xx} = 0.15E$ and $S_{yy} = 0.1E$, we used, $e^{[\frac{0.06a_e}{b_e} - 0.015]} - \frac{1}{a_e} = 0$ and $e^{[0.015 + \frac{0.09b_e}{a_e}]} - \frac{0.1}{b_e} = 0$. Whereas, when $S_{xx} = -0.15E$ and $S_{yy} = -0.1E$, we used, $e^{[\frac{-0.06a_e}{b_e} + 0.015]} - \frac{1}{a_e} = 0$ and $e^{[-\frac{0.09b_e}{a_e} - 0.015]} - \frac{0.1}{b_e} = 0$.

These four relations are plotted in Figure 11.7.1.

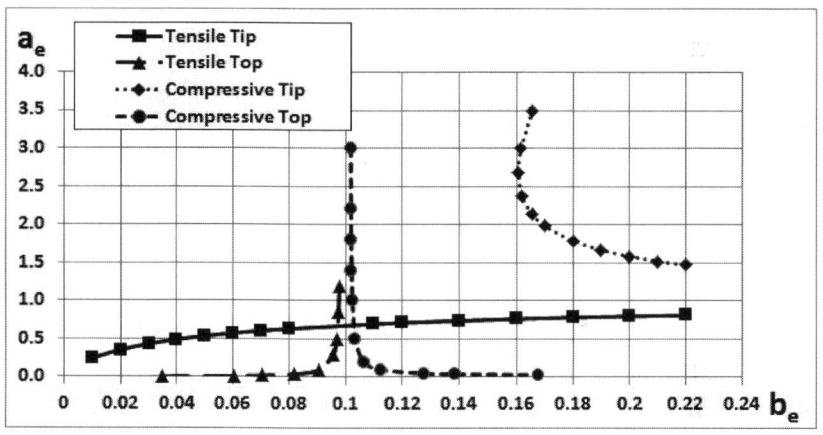

Figure 11.7.1: Graphs showing four logarithmic-Cauchy relations.

For logarithmic strain with Cauchy stress model and tensile loading, at the tip, we have $e^{[\frac{0.06a_e}{b_e} - 0.015]} - \frac{1}{a_e} = 0$. When

a_e is small, $\frac{1}{a_e} = e^{\left[\frac{0.06a_e}{b_e} - 0.015\right]}$ is large. Hence, b_e needs to be small. Thus, when $a_e = 0.2402$, b_e needs to be 0.01.

As a_e increases, $\frac{1}{a_e}$ decreases. Hence, $e^{\left[\frac{0.06a_e}{b_e} - 0.015\right]}$ needs to decrease and b_e needs to increase. A small increase in a_e needs a large increase in b_e. When $a_e = 1$, $b_e = 4$. In Figure 11.7.1, this relation is labelled "Tensile Tip".

At the top, we have $e^{\left[0.015 + \frac{0.09b_e}{a_e}\right]} - \frac{0.1}{b_e} = 0$. When b_e is small, $\frac{0.1}{b_e} = e^{\left[0.015 + \frac{0.09b_e}{a_e}\right]}$ is large. Hence, a_e needs to be small. Thus, when $b_e = 0.03474$, a_e needs to be 0.003.

As b_e increases, $\frac{0.1}{b_e}$ decreases. Hence, $e^{\left[0.015 + \frac{0.09b_e}{a_e}\right]}$ needs to decrease and a_e needs to increase. A small increase in b_e needs a large increase in a_e.

However, when $b_e = 0.1$, $\left[0.015 + \frac{0.09b_e}{a_e}\right]$ must equal 0. This is not possible since $a_e > 0$ and $b_e > 0$. Also, when $b_e > 0.1$, $\frac{0.1}{b_e}$ must be less than 1 and $\left[0.015 + \frac{0.09b_e}{a_e}\right]$ must be less than 0. This is also not possible. Consequently, $b_e < 0.1$. As b_e approaches 0.1 from left, a_e increases rapidly. In Figure 11.7.1, this relation is labelled "Tensile Top". As discussed in Chapter 10, "Tensile Tip" and "Tensile Top" in Figure 11.7.1 intersect at $a_e = 0.6709698413$ and $b_e = 0.097234704$.

For logarithmic strain with Cauchy stress model and compressive loading, at the tip, we have $e^{\left[\frac{-0.06a_e}{b_e} + 0.015\right]} - \frac{1}{a_e} = 0$.

When a_e is small, $\frac{1}{a_e} = e^{\left[\frac{-0.06a_e}{b_e} + 0.015\right]}$ is large. Hence, b_e needs to be large since the term with b_e in the denominator is negative. Thus, when $a_e = 0.9857$, b_e needs to be 100.0.

We note that since $a_e > 0$ and $b_e > 0$, the value of $e^{\left[\frac{-0.06a_e}{b_e}+0.015\right]}$ cannot be more than $e^{[0.015]} = 1.015113065$. Hence, $\frac{1}{a_e}$ cannot be more than 1.015113065. Consequently, a_e cannot be less than $\frac{1}{1.015113065} = 0.98511194$.

As a_e increases, $\frac{1}{a_e}$ decreases. Hence, $e^{\left[\frac{-0.06a_e}{b_e}+0.015\right]}$ needs to decrease, $\frac{-0.06a_e}{b_e}$ needs to increase, and b_e needs to decrease. When $a_e = 1$, $b_e = 4$. Thus, when a_e changes very little, from 0.9857 to 1, b_e changes from 100 to 4.

As the value of a_e increases, the need for b_e to decrease in order to increase $\frac{-0.06a_e}{b_e}$ reduces. Finally, the value of a_e is sufficient enough that b_e does not need to decrease such that b_e reaches its minimum value. Beyond this point, b_e increases to offset larger values of a_e.

In Figure 11.7.1, this relation is labelled "Compressive Tip". From this graph, we observe that as a_e increases b_e decreases until the point where b_e is slightly more than 0.16 then b_e increases. Above this minimum, a_e increases more rapidly than b_e.

To find the minimum value of b_e, we select a b_e value slightly greater than the expected value of the minimum. We select $b_e = 0.16067$. The vertical line at $b_e = 0.16067$ should intersect the curve at two points, one above the minimum and one below the minimum. By trial and error method, we find that these two points are $a_e = 2.6885655$ and $a_e = 2.6671297$ at $b_e = 0.16067$. Now, we find the midpoint of the chord joining these two points which is $a_e = 2.6778476$ and $b_e = 0.16067$.

The minimum value of b_e should be very close to the point of intersection of the line $a_e = 2.6778476$ and the curve $e^{[\frac{-0.06a_e}{b_e}+0.015]} - \frac{1}{a_e} = 0$. Substituting this value of a_e in the equation of the curve, we find that the point of intersection is $a_e = 2.6778476$ and $b_e = 0.1606687131$. Furthermore, at the minimum, $\frac{db_e}{da_e}$ should be zero. We use this condition to verify if this point of intersection is the minimum.

The equation of the curve is $e^{[\frac{-0.06a_e}{b_e}+0.015]} = \frac{1}{a_e}$. Or, $(a_e)e^{0.015} = e^{[\frac{0.06a_e}{b_e}]}$. Hence, differentiating, we obtain

$e^{0.015} = e^{[\frac{0.06a_e}{b_e}]}(0.06)\frac{d}{da_e}[\frac{a_e}{b_e}] = (0.06)e^{[\frac{0.06a_e}{b_e}]}[\frac{1}{b_e} - \frac{a_e}{b_e^2}\frac{db_e}{da_e}]$.

Therefore, $\frac{db_e}{da_e} = (\frac{b_e^2}{a_e})[\frac{1}{b_e} - (\frac{1}{0.06})e^{[0.015-\frac{0.06a_e}{b_e}]}]$.

At the minimum, $\frac{db_e}{da_e} = 0$ requires that

$[\frac{1}{b_e} - (\frac{1}{0.06})e^{[0.015-\frac{0.06a_e}{b_e}]}] = 0$.

Or, $[(\frac{b_e}{0.06})e^{[0.015-\frac{0.06a_e}{b_e}]}] - 1 = 0$.

At the point of intersection, $a_e = 2.6778476$ and $b_e = 0.1606687131$,

$[(\frac{0.1606687131}{0.06})e^{[0.015-\frac{(0.06)2.6778476}{0.1606687131}]}] - 1 = -0.000013337$.

Thus, the minimum is just left of $b_e = 0.1606687131$. By reducing b_e to 0.16066871305, we find another point on the curve where $a_e = 2.677812$ and

$[(\frac{0.16066871305}{0.06})e^{[0.015-\frac{(0.06)2.677812}{0.16066871305}]}] - 1 = -0.000000043$. This point is very close to the minimum. It is not necessary to find the exact minimum.

I believe, the presence of this minimum and the nature of this relation at the tip are unique to the use of logarithmic strain definition for compressive loading.

For logarithmic strain with Cauchy stress model and compressive loading, at the top, we have $e^{[-\frac{0.09b_e}{a_e} - 0.015]} - \frac{0.1}{b_e} = 0$. This equation can be rewritten as

$$e^{[-\frac{0.09b_e}{a_e} - 0.015]} = \frac{0.1}{b_e} = \frac{1}{e^{[\frac{0.09b_e}{a_e} + 0.015]}}.$$

Or, $(0.1)e^{[\frac{0.09b_e}{a_e} + 0.015]} = b_e$

When $b_e = 0.1$, $[0.015 + \frac{0.09b_e}{a_e}]$ must equal 0. This is not possible since $a_e > 0$ and $b_e > 0$. Also, when $b_e < 0.1$, $\frac{b_e}{0.1}$ must be less than 1 and $[0.015 + \frac{0.09b_e}{a_e}]$ must be less than 0. This is also not possible. Consequently, $b_e > 0.1$.

When a_e is very large, as compared to b_e, $\frac{0.09b_e}{a_e}$ approaches 0, $e^{[\frac{0.09b_e}{a_e} + 0.015]}$ approaches $e^{[0.015]} = 1.015113$, and b_e approaches 0.1015113. As b_e increases beyond $b_e = 0.1015113$, a_e reduces rapidly. Thus, when $b_e = 0.101822$, $a_e = 3.0$ and when $b_e = 0.102452$, $a_e = 1.0$. However, as b_e increases beyond $b_e = 0.12$, a_e reduces very slowly. In Figure 11.7.1, this relation is labelled "Compressive Top".

The above analysis reveals that relations Compressive Tip and Compressive Top for logarithmic strain with Cauchy stress do not intersect. Out of the four relations shown in Figure 11.7.1, only the relation Compressive Tip displays a b_e minimum. This minimum occurs before the relation Compressive Tip crosses the relation Compressive Top.

We observed a similar situation in Figure 10.2.6 for model

with logarithmic strain and Kirchoff stress where applied loads were $S_{xx} = -0.15E$ and $S_{yy} = -0.1E$.

So, can we conclude that models with logarithmic strain definition are not appropriate for compressive loads? For this purpose, several scenarios were considered. Below, we compare the the model with logarithmic strain and Cauchy stress using tensile loads and compressive loads as above. However, we change the size of the initial opening to $a_s = 1$ and $b_s = 0.9$ instead of $b_s = 0.1$.

Since we change only b_s, only two relations at the top are affected. Now, when $S_{xx} = 0.15E$ and $S_{yy} = 0.1E$, we use, $e^{[\frac{0.06a_e}{b_e} - 0.015]} - \frac{1}{a_e} = 0$ and $e^{[0.015 + \frac{0.09b_e}{a_e}]} - \frac{0.9}{b_e} = 0$. Whereas, when $S_{xx} = -0.15E$ and $S_{yy} = -0.1E$, we use, $e^{[\frac{-0.06a_e}{b_e} + 0.015]} - \frac{1}{a_e} = 0$ and $e^{[-\frac{0.09b_e}{a_e} - 0.015]} - \frac{0.9}{b_e} = 0$.

These four relations are plotted in Figure 11.7.2.

We note that in Figure 11.7.1, b_e values range from 0 to 0.24 whereas in Figure 11.7.2 b_e values range from 0 to 4.5. Graphs for relations at the tip remain the same. However, graphs for relations at the top are shifted to right. Since $b_s = 0.9$, for the relation Tensile Top, $b_e < 0.9$. And for the relation Compressive Top, $b_e > 0.9$. As a result, relations for Compressive Tip and Compressive Top now intersect to yield a solution. The solution for tensile loads is $a_e = 0.94715465615$ and $b_e = 0.82013149495$. The solution for compressive loads is $a_e = 1.0494707621$ and $b_e = 0.9949790588$.

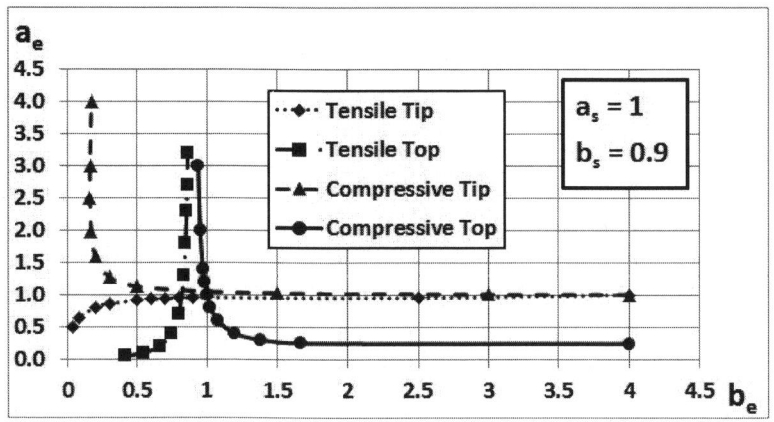

Figure 11.7.2: Graphs showing four logarithmic-Cauchy relations.

In this book, we mostly considered four different load conditions when $a_s = 1$ and $b_s = 0.1$ such that $\frac{b_s}{a_s} = 0.1$. Only in this section, we changed $b_s = 0.1$ to $b_s = 0.9$. This change in initial geometry provided us a solution where there was none. This illustrates that the initial geometry also plays a significant role.

Thus, incompatibility of mathematical relations used in defining strain with loading conditions and/or initial geometry can be a limitation.

Chapter 12

Summary and Conclusions

On page 124 of the first book (titled ***Understanding the Elastic Stress Field Around an Elliptical Hole in a Thin Plate (in the deformed configuration)-ISBN 978-1-48359-262-6***), I had mentioned :

"(1) It will be shown that the Singh, Glinka, and Dubey (1994) [1] *finding is a special case of the general solution developed in Phase 2.*

(2) Two material properties, Young's modulus and Poisson's ratio affect the final dimensions (not just the Young's modulus)."

---

[1]M. Singh, G. Glinka, and R. Dubey
Notch and crack analysis as a moving boundary problem.
Engineering Fracture Mechanics, 47(4):479–492, 1994.

On page 74 of this second book (a product of Phase 2 of this research), we noted that if we let $\nu = 1$, in (8.3.2), we obtain (8.4.1). Thus, the expression developed by Singh, Glinka, and Dubey is a special case of the general solution developed here. Consequently, the expression developed by Singh, Glinka, and Dubey is applicable only to materials with Poisson's ratio equal to 1.

Furthermore, all expressions developed for all models in Chapter 8 and Chapter 9 of this book contain E, Young's Modulus. and ν, Poisson's Ratio.

Based on the detailed analysis carried out in this book, summary and conclusions from various chapters are presented below.

Chapter 1 : Cartesian co-ordinate system and Elliptical co-ordinate system are compared in this Chapter. In Cartesian co-ordinate system, we need any two of the three i) a, ii) b, and iii) c, to define an ellipse since we have one relation, $a^2 - b^2 = c^2$. In Elliptical co-ordinate system, we need any two of the four i) a, ii) b, iii) c, and iv) α, to define an ellipse since we have two relations, $a = c \cosh\alpha$ and $b = c \sinh\alpha$.

In Cartesian co-ordinate system, $x = $ constant is a vertical line. Similarly, in Elliptical co-ordinate system, $\alpha = $ constant is an ellipse. However, $x = $ constant is a specific line whereas $\alpha = $ constant is not a specific ellipse. In addition to α, we need c, to define a specific ellipse.

In Cartesian co-ordinate system, we need x and y to define a point. In Elliptical co-ordinate system, we need α, β, and c to define a point. Since $x = c \cosh\alpha \cos\beta$ and

$y = c \sinh\alpha \sin\beta$, the same (α, β) point can be associated with many (x, y) points, depending on chosen values of c. Thus, there is no one to one correspondence between (α, β) co-ordinates and (x,y) co-ordinates.

When c decreases, α curves (ellipses) and β curves (hyperbolas) move towards the center. Conversely, when c increases, α curves and β curves move away from the center. In Cartesian system the x-y grid is fixed, but in Elliptical system the α-β grid is not fixed; it is linked to c. When $c = 0$, focal points and center merge, ellipses become circles, and vertices of hyperbolas merge into the origin. Hyperbolas become straight lines (radii) converging at the center. Elliptical co-ordinate system becomes Polar co-ordinate system.

For all values of c, when α is very large, $\sinh\alpha \approx \cosh\alpha$ such that the Elliptical system approximates the Polar system, far away from the center.

If a material point moves due to an increase in c only, without any change in α and β, then the point moves away from the center only because the grid has expanded due to an increase in c. Hence, the (α, β) co-ordinates of the point do not change although (x, y) co-ordinates change. On the other hand, if a material point moves due to an increase in α only, without any change in c and β, then the point moves away from the center along a β curve. Since c does not change, the grid is fixed. Hence, only the (α) co-ordinate of the point changes and (x, y) co-ordinates change.

Chapter 2 : Many cases were studied although the detailed analysis for the case where α increases and c decreases was presented in Chapter 2. This analysis of displacements at various points along the surface of an elliptical hole revealed the following :

1) At the tip, where $\beta = 0$, the displacement is always horizontal and perpendicular to the contour, whereas at the top, where $\beta = \frac{\pi}{2}$, the displacement is always vertical and perpendicular to the contour. Various studies by some researchers have used this feature on the basis of symmetry although they have not used the Elliptical co-ordinate system.

2) Along other points on the contour, the displacement may not be perpendicular to the contour. Some researchers have misunderstood this feature.

3) Displacements can be broken down in to two components, due to a change in α and due to a change in c.

Chapter 3 : As mentioned above, when $c = 0$, Elliptical co-ordinate system becomes Polar co-ordinate system. For Polar co-ordinate system, expression for r in terms of x and y as well as expression for θ in terms of x and y are developed in Chapter 3.

For Elliptical co-ordinate system, the following six expressions are developed in Chapter 3. They are 1) expression for c in terms of x, y, and α, 2) expression for c in terms of x, y, and β, 3) expression for α in terms of x, y, and c, 4) expression for α in terms of x, y, and β, 5) expression for β in terms of x, y, and c, and 6) expression for β in terms of x, y, and α.

In addition, I have also presented descriptions of the nature of f_1, f_2, and f_3.

Chapter 4 : We note that material points along $\beta = 0$ before deformation remain along $\beta = 0$ even after deformation. Material points along $\beta = \frac{\pi}{2}$ before deformation remain along $\beta = \frac{\pi}{2}$ even after deformation. Similarly, material points along a β curve between $\beta - 0$ and $\beta = \frac{\pi}{2}$ remain on that β curve.

Thus, material points on each elliptical contour do not jump around. Each elliptical contour may change its shape and/or size due to changes in a, b, or both. As a consequence, the locations of points and the spacing between points change but the order in which they are held on an elliptical contour remains the same. Thus, β curves may shift along with material points on them as c changes. Yet, β curves remain in the same order. Even when α value changes, all material points on an elliptical contour move towards or away from the center while they remain on the same β curves. As a result, although c and α values change, β values associated with material points are not affected.

Using relevant expressions from Chapter 3, expressions for $\frac{\partial c}{\partial x}$, $\frac{\partial c}{\partial y}$, $\frac{\partial \alpha}{\partial x}$, and $\frac{\partial \alpha}{\partial y}$ are developed in Chapter 4.

Chapter 5 : The displacement of a point from starting position (x_s, y_s) to ending position (x_e, y_e) is u which has two components, u_x in direction x and u_y in direction y. We know $u_x = x_e - x_s$ such that $\frac{\partial u_x}{\partial x_s} = \frac{\partial x_e}{\partial x_s} - 1$.

Furthermore, u_x is made up of two parts : 1) due to change in only α and 2) due to change in only c. Thus,

$$u_x = (x_e - x_s)_{change\ in\ only\ \alpha} + (x_e - x_s)_{change\ in\ only\ c}$$

$$= c_e \cos\beta(\cosh\alpha_e - \cosh\alpha_s) + \cosh\alpha_s \cos\beta(c_e - c_s)$$

Expressions for $\frac{\partial u_x}{\partial x_s}$ and $\frac{\partial x_e}{\partial x_s}$ are developed by differentiating this expression for u_x with respect to x_s.

Similarly, expressions for $\frac{\partial u_x}{\partial y_s}$, $\frac{\partial x_e}{\partial y_s}$, $\frac{\partial u_y}{\partial x_s}$, $\frac{\partial y_e}{\partial x_s}$, $\frac{\partial u_y}{\partial y_s}$, and $\frac{\partial y_e}{\partial y_s}$ are developed in Chapter 5.

Using these expressions, it is shown that, at the tip,

$\frac{\partial x_e}{\partial x_s} = \frac{a_e}{a_s}$, $\frac{\partial u_x}{\partial x_s} = \frac{a_e - a_s}{a_s}$, $\frac{\partial u_x}{\partial y_s} = \frac{\partial x_e}{\partial y_s} = 0$,

$\frac{\partial u_y}{\partial x_s} = \frac{\partial y_e}{\partial x_s} = 0$, $\frac{\partial y_e}{\partial y_s} = 1$ and $\frac{\partial u_y}{\partial y_s} = \frac{\partial y_e}{\partial y_s} - 1 = 0$.

and, at the top,

$\frac{\partial x_e}{\partial x_s} = 1$, $\frac{\partial u_x}{\partial x_s} = 0$, $\frac{\partial u_x}{\partial y_s} = \frac{\partial x_e}{\partial y_s} = 0$,

$\frac{\partial u_y}{\partial x_s} = \frac{\partial y_e}{\partial x_s} = 0$, $\frac{\partial y_e}{\partial y_s} = \frac{b_e}{b_s}$ and $\frac{\partial u_y}{\partial y_s} = \frac{\partial y_e}{\partial y_s} - 1 = \frac{b_e - b_s}{b_s}$.

Chapter 6 : In the first book (titled ***Understanding the Elastic Stress Field Around an Elliptical Hole in a Thin Plate (in the deformed configuration) ISBN 978-1-48359-262-6***), I had mentioned in Chapter 9, "The Phase 2 of this research, to be presented in the near future, accepts results from the stress analysis and carries them forward with the help of linear stress-strain relations and four definitions of strain....."

The Phase 1 of this research allowed us to study exclusively the impact of geometric non-linearity due to the presence of an elliptical hole through stress analysis. In this Phase 2, we developed strain-displacement relations in Elliptical co-ordinate system in the previous five chapters.

Now, we employ linear stress-strain relations to link these two modules. Subsequently, we apply boundary conditions to arrive at the final configuration of the elliptical hole .

The stress-strain relations in two dimensions are

$e_{xx} = \frac{1}{E}[\sigma_{xx} - \nu\sigma_{yy}]$, $e_{yy} = \frac{1}{E}[\sigma_{yy} - \nu\sigma_{xx}]$, and

$e_{xy} = \frac{1+\nu}{E}[\sigma_{xy}] = e_{yx} = \frac{1+\nu}{E}[\sigma_{yx}]$

Boundary conditions are :

At the tip, $\sigma_{xx} = \sigma_{xy} = \sigma_{yx} = 0$ and

$$\sigma_{yy} = [S_{yy}][1 + (\frac{2a_e}{b_e})] - [S_{xx}].$$

And, at the top, $\sigma_{yy} = \sigma_{xy} = \sigma_{yx} = 0$ and

$$\sigma_{xx} = [S_{xx}][1 + (\frac{2b_e}{a_e})] - [S_{yy}].$$

Hence, substituting these expressions for stresses in stress-strain relations,

at the tip, $e_{xx} = \frac{-\nu}{E}[S_{yy}(1 + \frac{2a_e}{b_e}) - S_{xx}]$,

$$e_{yy} = \frac{1}{E}[S_{yy}(1 + \frac{2a_e}{b_e}) - S_{xx}], \text{ and}$$

$$e_{xy} = \frac{1+\nu}{E}[\sigma_{xy}] = e_{yx} = \frac{1+\nu}{E}[\sigma_{yx}] = 0.$$

And, at the top, $e_{xx} = \frac{1}{E}[S_{xx}(1 + \frac{2b_e}{a_e}) - S_{yy}]$,

$$e_{yy} = \frac{-\nu}{E}[S_{xx}(1 + \frac{2b_e}{a_e}) - S_{yy}], \text{ and}$$

$$e_{xy} = \frac{1+\nu}{E}[\sigma_{xy}] = e_{yx} = \frac{1+\nu}{E}[\sigma_{yx}] = 0.$$

Chapter 7 : I have intentionally deferred this topic (definitions of strain) as much as possible so that matters of choice are left towards the end. Indeed, what definition of strain a researcher or an analyst should use is his/her choice.

I recall what a senior researcher once noted during a conversation. He implied that materials do not worry about what definition of strain we use. Materials simply deform due to stress. It is our job to select appropriate strain definition. In addition to the expected magnitudes of induced stresses and strains, one should also consider under what conditions the parameters reflecting the properties of material were evaluated, initial geometry of the hole, as well as far field applied loads.

Here, my objective is not to discuss criteria for the selection or pros and cons of different strain definitions. My goal is to develop formulas for each definition of strain so that researchers and analysts have access to these formulas and can readily make appropriate selections. And, if needed, one can use more than one definition to evaluate different "what if" scenarios.

Four definitions of strain are presented in Chapter 7. They are for 1) Engineering strain, 2) Green strain, 3) Logarithmic strain, and 4) Almansi strain.

I want to emphasize that so far we have made no assumption about the magnitude of strain. Furthermore, three of these four definitions are non-linear.

Chapter 8 : This chapter is exclusively devoted to the engineering strain which is the only linear definition of strain in this book. Even while dealing with the engineering strain, we have not made the small strain assumption.

At the tip, displacement is only in horizontal direction, from a_s to a_e. Hence, $\lambda = \frac{a_e}{a_s}$ and

$$e_{xx} = e_{eng} = \lambda - 1 = \frac{a_e}{a_s} - 1 = \frac{-\nu}{E}[S_{yy}(1 + \frac{2a_e}{b_e}) - S_{xx}].$$

Similarly, at the top, displacement is only in vertical direction, from b_s to b_e. Hence, $\lambda = \frac{b_e}{b_s}$ and

$$e_{yy} = e_{eng} = \lambda - 1 = \frac{b_e}{b_s} - 1 = \frac{-\nu}{E}[S_{xx}(1 + \frac{2b_e}{a_e}) - S_{yy}].$$

Solving these two equations for a_e and b_e, as shown in Appendix E, we obtain

$$a_e = a_s b_s \left[\frac{(1 - \frac{\nu}{E}S_{xx} - \frac{\nu}{E}S_{yy})(1 + \frac{\nu}{E}S_{xx} + \frac{\nu}{E}S_{yy})}{b_s(1 + \frac{\nu}{E}S_{yy} - \frac{\nu}{E}S_{xx}) + 2a_s(\frac{\nu}{E}S_{yy})}\right] \text{ and}$$

$$b_e = a_s b_s \left[\frac{(1 - \frac{\nu}{E}S_{xx} - \frac{\nu}{E}S_{yy})(1 + \frac{\nu}{E}S_{xx} + \frac{\nu}{E}S_{yy})}{a_s(1 + \frac{\nu}{E}S_{xx} - \frac{\nu}{E}S_{yy}) + 2b_s(\frac{\nu}{E}S_{xx})}\right].$$

From the above two expressions we conclude that 1) when $a_s = 0$ or $b_s = 0$, both a_e and b_e vanish. Hence, a_s and b_s both must be greater than zero, 2) when $S_{xx} = 0$, b_e is not affected by a_s, and 3) when $S_{yy} = 0$, a_e is not affected by b_s.

It is shown that a circular hole under uniaxial stress cannot remain a circular hole while an elliptical hole under uniaxial stress may remain an elliptical hole, although its focal length may change.

Some conditions under which an elliptical hole under uniaxial tension can deform to a circular hole are discussed.

We know at the tip, $\sigma_{yy} = [S_{yy}][1 + (\frac{2a_e}{b_e})] - [S_{xx}]$.

Substituting values for a_e and b_e from above in this expression, as shown in Appendix E, we obtain

$$\sigma_{yy} = S_{yy} - S_{xx} + 2S_{yy}\left[\frac{a_s(1 + \frac{\nu}{E}S_{xx} - \frac{\nu}{E}S_{yy}) + 2b_s(\frac{\nu}{E}S_{xx})}{b_s(1 + \frac{\nu}{E}S_{yy} - \frac{\nu}{E}S_{xx}) + 2a_s(\frac{\nu}{E}S_{yy})}\right].$$

When $S_{xx} = 0$, the above expression simplifies to

$$\sigma_{yy} = S_{yy} \left[\frac{b_s(1 + \frac{\nu}{E}S_{yy}) + 2a_s}{b_s(1 + \frac{\nu}{E}S_{yy}) + 2a_s(\frac{\nu}{E}S_{yy})} \right]$$

Consequently, if the far field applied stress is only in the direction perpendicular to the major axis of the hole then the stress concentration, at the tip, is

$$\frac{\sigma_{yy}^{tip}}{S_{yy}} = \left[\frac{1 + (1 + \frac{\nu}{E}S_{yy})\frac{b_s}{2a_s}}{(\frac{\nu}{E}S_{yy}) + \frac{b_s}{2a_s}(1 + \frac{\nu}{E}S_{yy})} \right].$$

If we let $\nu = 1$ in this last expression, we obtain

$$\frac{\sigma_{yy}^{tip}}{S_{yy}} = \left[\frac{1 + (1 + \frac{S_{yy}}{E})\frac{b_s}{2a_s}}{(\frac{S_{yy}}{E}) + \frac{b_s}{2a_s}(1 + \frac{S_{yy}}{E})} \right].$$

We now compare this expression with a similar expression developed by Singh, Glinka, and Dubey (see expression (37) on page 485 of their paper) $\frac{\sigma_{22}^A}{S} = \frac{1 + [1 + (S/E)]\frac{b_i}{2a_i}}{S/E + (1 + (S/E))\frac{b_i}{2a_i}}.$

These two expressions are identical. Thus, the expression developed by Singh, Glinka, and Dubey is a special case of the general solution developed here. Consequently, the expression developed by Singh, Glinka, and Dubey is applicable only to materials with Poisson's ratio equal to 1.

Finally, we note that all expressions developed in this chapter contain E and ν whereas work by Singh, Glinka, and Dubey does not contain ν.

If we let $b_s = 0$ in $\frac{\sigma_{yy}^{tip}}{S_{yy}} = \left[\frac{1 + (1 + \frac{\nu}{E}S_{yy})\frac{b_s}{2a_s}}{(\frac{\nu}{E}S_{yy}) + \frac{b_s}{2a_s}(1 + \frac{\nu}{E}S_{yy})} \right]$ then it reduces to $\frac{\sigma_{yy}^{tip}}{S_{yy}} = \left[\frac{1+0}{(\frac{\nu}{E}S_{yy})+0}\right] = \frac{E}{\nu}\frac{1}{S_{yy}}.$

Or, when $b_s = 0$ and $S_{yy} > 0$, $\sigma_{yy}^{tip} = \frac{E}{\nu}$.

Similarly, if we let $b_i = 0$ in $\frac{\sigma_{22}^A}{S} = \frac{1+[1+(S/E)]\frac{b_i}{2a_i}}{S/E+(1+(S/E))\frac{b_i}{2a_i}}$,

it reduces to $\frac{\sigma_{22}^A}{S} = \frac{1+0}{S/E+0} = \frac{E}{S}.$

Or, when $b_i = 0$ and $S > 0$, $\sigma_{22}^A = E$.

Hence, Singh, Glinka, and Dubey (see expression (44) and the subsequent paragraph on page 486 of their paper) stated, *"It is interesting to note that, contrary to the classical solution, the stress at the crack tip is finite. Moreover, according to eq. (44), the stress in the crack tip remains constant and equal to the modulus of elasticity, E, regardless of the applied load S and the initial length a_i."*

On the other hand, as discussed in Appendix E, even if two surfaces are closely touching each other, b_s exists, however small it may be. We can argue that if $b_s = 0$ then there is no crack. Hence, $b_s > 0$. Thus, for sharp cracks where b_s is much smaller than a_s, the tip stress approaches $\frac{E}{\nu}$.

Expressions derived in this chapter are compared with Singh, Glinka, and Dubey findings as well as small deformation theory in detail in Chapter 8.

I would like to emphasize that however small in magnitude, due to the presence of far field stresses, the deformations must occur. Therefore, $a_e \neq a_s$ and $b_e \neq b_s$. Hence, the values obtained by using the small deformation theory are not as reliable. On the other hand, they provide upper bounds to σ_{yy}^{tip}/E values.

Chapter 9 : In Chapter 8, for the model with engineering strain and Cauchy stress, we developed two relations between a_e and b_e with the help of two boundary conditions - one at the tip and the other at the top.

In Chapter 9, we use three more strain definitions and two stress definitions to develop similar relations to compute a_e and b_e for five different models. They are : (1) Green strain with Cauchy stress, (2) Logarithmic strain with Cauchy stress, (3) Logarithmic strain with Kirchoff stress, (4) Almansi strain with Cauchy stress, and (5) Almansi strain with Kirchoff stress.

My goal is to illustrate that to study an elliptical hole in a thin plate, an analyst can select various combinations of definitions of stresses and strains and arrive at better understanding through the analytical tools developed here. Finally. based on such an understanding, an analyst can make a well educated decision.

Experimental work can help us decide which model is best suited for which materials under what conditions and possibly why.

There are many textbooks that describe various definitions of stresses, definitions of strains, and relations among these definitions. However, I recommend the following two references :

(1) *https://www.comsol.com/blogs/why-all -these-stresses-and-strains/ written by Henrik Sönnerlind*

and

(2) *http://solidmechanics.org/Text/Chapter5_3/ Chapter5_3.php written by Allan F. Bower.*

The first two paragraphs in Chapter 7 (of the second reference), *"Introduction to Finite Element Analysis in Solid Mechanics"*, describe the need and importance as well as concerns regarding the use of commercial finite element software programs.

Also, section 5.3.2, *"Demonstration that the complex variable solution satisfies the governing equations"* is also relevant to our discussions here.

Kanezaki, Nagata, and Murakami [2007] [2] developed a closed-form solution for stress distribution around an elliptical hole. Their solution is in Cartesian co-ordinate system. However, they do not use the concept of the deformed configuration. Hence, their expressions for stress components do not contain any variables or parameters that relate to the deformed configuration.

In contrast, expressions for stress components developed in my first book contain the parameter α_e that relates to the deformed configuration. Furthermore, in this second book we developed six models to determine the shape and size of the deformed hole.

---

[2]Toshihiko Kanezaki, Koichi Nagata, and Yukitaka Murakami. New closed-form solution by cartesian co-ordinate for stress distribution around elliptic hole and its applications. *Journal of Solid Mechanics and Materials Engineering*, 1:232–243, 2007.

Chapter 10 : In this chapter, procedures to compute a_e and b_e for five of the six models (except for the model with engineering strain which was studied in Chapter 8) are developed. For each model, initial size of the opening and Poisson's ratio are set at $a_s = 1$, $b_s = 0.1$, and $\nu = 0.3$. For each model, four different loading conditions are used. They are 1) $\frac{S_{xx}}{E} = 0$, and $\frac{S_{yy}}{E} = 0.1$, 2) $\frac{S_{xx}}{E} = 0.1$, and $\frac{S_{yy}}{E} = 0$, 3) $\frac{S_{xx}}{E} = 0.15$, and $\frac{S_{yy}}{E} = 0.1$, and 4) $\frac{S_{xx}}{E} = -0.15$, and $\frac{S_{yy}}{E} = -0.1$. Thus, computational procedures for twenty different situations are examined. For each situation, there are two unknowns a_e and b_e while there are two simultaneous equations - one to satisfy the boundary condition at the tip and the other to satisfy the boundary condition at the top.

Three procedures are used to solve these equations. Procedure 1 is used when it is possible to isolate one of the two unknowns in one of the two simultaneous equations. For example, for the model with Green strain, when $S_{yy} = 0$, as shown in Appendix F, we obtain $a_e = a_s\sqrt{1 + 2\nu\left(\frac{S_{xx}}{E}\right)}$. Once we compute one such unknown, the other simultaneous equation with the other unknown becomes easier to solve.

Procedure 2 is used when it is possible to have only one unknown in one of the two simultaneous equations. But, it is not possible to isolate it since it appears in two or more terms which cannot be combined. For example, for the model with Green strain, when $\frac{S_{xx}}{E} = 0.15$, and $\frac{S_{yy}}{E} = 0.1$, we obtain equation (F.0.1), $\dfrac{(b_e)^2(0.0018)}{-0.06 + \sqrt{0.0036 + (b_e)^2 1.03}} + (b_e)^2 -$

$0.0097 = 0$. Such an equation can be solved by trial and error method, see Table 10.1.1.

Procedure 3 is used when it is not possible to have only one unknown in any one of the two simultaneous equations. Thus, two unknowns a_e and b_e appear in both equations. For example, for the model with logarithmic strain and Cauchy stress, when $\frac{S_{xx}}{E} = 0.15$, and $\frac{S_{yy}}{E} - 0.1$, we obtain equation (10.2.1) $e^{\left[\frac{0.06a_e}{b_e} - 0.015\right]} - \frac{1}{a_e} = 0$ and equation (10.2.2) $e^{\left[0.015 + \frac{0.09b_e}{a_e}\right]} - \frac{0.1}{b_e} = 0$. Trial and error method described in detail in Appendix G is employed to obtain five solutions that satisfy equation (10.2.1) and five solutions that satisfy equation (10.2.2). These ten solutions are plotted in Figure 10.2.1 showing two curves. The point of intersection of these two curves is the solution we are seeking.

For the model with engineering strain, it was possible to isolate both a_e in equation (8.2.2) and and b_e in equation (8.2.3). Hence, the first row in Table 12.0.1 for engineering strain shows for all four loading conditions that it does not matter whether a_e is computed first or b_e.

For the model with Green strain, in section 9.1, we developed equation (9.1.4) where only unknown was b_e (we could have developed an equation where only unknown was a_e). Furthermore, when $S_{xx} = 0$ or $S_{yy} = 0$, equation (9.1.4) reduces such that b_e can be isolated and procedure 1 can be used. Subsequently, a_e can be computed using procedure 1. Therefore, _1b then 1a_ appear in the first two columns. However, when neither $S_{xx} = 0$ nor $S_{yy} = 0$, equation (9.1.4) cannot be reduced. We need procedure 2 to determine b_e.

Load → / Model ↓	Initial Geometry $a_s = 1$ and $b_s = 0.1$ / Poisson's Ratio $\nu = 0.3$			
	$S_{xx} = 0,$ $S_{yy} = 0.1E$	$S_{xx} = 0.1E,$ $S_{yy} = 0$	$S_{xx} = 0.15E,$ $S_{yy} = 0.1E$	$S_{xx} = -0.15E,$ $S_{yy} = -0.1E$
Engineering strain Cauchy stress	1a or 1b then 1b or 1a	1a or 1b then 1b or 1a	1a or 1b then 1b or 1a	1a or 1b then 1b or 1a
Green strain Cauchy stress	1b then 1a	1b then 1a	2b then 1a	2b then 1a
Logarithmic strain Cauchy stress	1b then 2a	1a then 2b	3	No solution No solution
Logarithmic strain Kirchoff stress	2b then 2a	2a then 2b	3	No solution No solution
Almansi strain Cauchy stress	1b then 2a	1a then 2b	2a then 1b, or 2b then 1a, or 3	2a then 1b, or 2b then 1a, or 3
Almansi strain Kirchoff stress	2b then 2a	2a then 2b	2a then 1b, or 2b then 1a, or 3	No solution No solution

Table 12.0.1: Table showing through which procedure and in which order a_e and b_e for each loading condition for each of the six models were computed.

Subsequently, a_e can be computed using procedure 1. Therefore, <u>2b then 1a</u> appear in the last two columns.

For the model with logarithmic strain and Cauchy stress, when $S_{xx} = 0$, equation (9.2.2) simplifies such that b_e can be isolated and procedure 1 can be used. Subsequently, a_e can be computed using procedure 2. Therefore, <u>1b then 2a</u> appear in the first column. When $S_{yy} = 0$, equation (9.2.1) simplifies such that a_e can be isolated and procedure 1 can be used. Subsequently, b_e can be computed using procedure 2. Therefore, <u>1a then 2b</u> appear in the second column. However, when neither $S_{xx} = 0$ nor $S_{yy} = 0$, neither equation (9.2.1) nor equation (9.2.2) can be reduced. When $S_{xx} > 0$ and $S_{yy} > 0$, we need procedure 3 to determine a_e and b_e. However, when $S_{xx} < 0$ and $S_{yy} < 0$, there is no solution.

For the model with logarithmic strain and Kirchoff stress, when $S_{xx} = 0$, equation (9.2.4) simplifies. But, b_e appears in more than one term. Hence, procedure 2 can be used. Subsequently, a_e can be computed using procedure 2. Therefore, <u>2b then 2a</u> appear in the first column. When $S_{yy} = 0$, equation (9.2.3) simplifies. But, a_e appears in more than one term. Hence, procedure 2 can be used. Subsequently, b_e can be computed using procedure 2. Therefore, <u>2a then 2b</u> appear in the second column. However, when neither $S_{xx} = 0$ nor $S_{yy} = 0$, neither equation (9.2.3) nor equation (9.2.4) can be reduced. When $S_{xx} > 0$ and $S_{yy} > 0$, we need procedure 3 to determine a_e and b_e. However, when $S_{xx} < 0$ and $S_{yy} < 0$, there is no solution.

For the model with Almansi strain and Cauchy stress, the behaviour is similar to the model with logarithmic strain

181

and Cauchy stress when $S_{xx} = 0$ (i.e. procedure $\underline{1b \ then \ 2a}$) and $S_{yy} = 0$ (i.e. procedure $\underline{1a \ then \ 2b}$). However, when $S_{xx} > 0$ and $S_{yy} > 0$, it is possible to express b_e in terms of a_e. Thus, equation (10.3.1) can be rewritten as $\left[\frac{0.12}{b_e}\right]a_e^3 = 1 - \left[0.97\right]a_e^2$. Or, $b_e = \left[\frac{0.12a_e^3}{1-0.97a_e^2}\right]$. Substituting this expression for b_e in equation (10.3.2), we obtain $\left[\frac{18}{a_e}\right]b_e^3 + \left[103\right]b_e^2 - 1 = \left[\frac{18}{a_e}\right]\left[\frac{0.12a_e^3}{1-0.97a_e^2}\right]^3 + \left[103\right]\left[\frac{0.12a_e^3}{1-0.97a_e^2}\right]^2 - 1 = 0$. This equation contains only one unknown a_e. It can be solved by procedure 2. We can verify that $a_e = 0.73113058511$ is a solution to this equation. Subsequently, we can find $b_e = \left[\frac{0.12a_e^3}{1-0.97a_e^2}\right]$ using procedure 1. Thus, we can solve equation (10.3.1) and equation (10.3.2) using procedure $\underline{2a \ then \ 1b}$. If we use equation (10.3.2) to obtain $a_e = \left[\frac{18b_e^3}{1-103b_e^2}\right]$, we can find b_e first and then a_e. Thus, we can solve equation (10.3.1) and equation (10.3.2) in three ways - procedure $\underline{2a \ then \ 1b}$ or procedure $\underline{2b \ then \ 1a}$ or procedure $\underline{3}$. Similarly, when $S_{xx} < 0$ and $S_{yy} < 0$, $\left[\frac{-0.12}{b_e}\right]a_e^3 + \left[1.03\right]a_e^2 - 1 = 0$, at the tip and $-\left[\frac{0.18}{a_e}\right]b_e^3 + \left[0.97\right]b_e^2 - 0.01 = 0$, at the top can be solved in three ways - procedure $\underline{2a \ then \ 1b}$ or procedure $\underline{2b \ then \ 1a}$ or procedure $\underline{3}$.

For the model with Almansi strain and Kirchoff stress, the behaviour is similar to the model with logarithmic strain and Kirchoff stress when $S_{xx} = 0$ (i.e. procedure $\underline{2b \ then \ 2a}$) and $S_{yy} = 0$ (i.e. procedure $\underline{2a \ then \ 2b}$). When $S_{xx} > 0$ and $S_{yy} > 0$, as in the model with Almansi strain and Cauchy stress, we can solve equation (10.3.3) and equation (10.3.4) each containing two unknowns, a_e and b_e, in three ways - procedure _2a then 1b_ or procedure _2b then 1a_ or procedure _3_. However, when $S_{xx} < 0$ and $S_{yy} < 0$, as discussed in Chapter 10, there is no solution.

Chapter 11 : First, in order to complete a comparison for the four loading conditions, we compute a_e and b_e for four loading conditions for the model with engineering strain. Next, for an easy comparison, we present a_e and b_e for each of the six models and each of the four loading conditions (i.e. 24 situations) in Table 11.2.1.

We note that when $\frac{S_{xx}}{E} = 0$ and $\frac{S_{yy}}{E} = 0.1$, b_e values are not too different, whereas a_e values exhibit a significant variation. However, when $\frac{S_{xx}}{E} = 0.1$ and $\frac{S_{yy}}{E} = 0$, b_e values as well as a_e values do not show much variation.

When $\frac{S_{xx}}{E} = 0.15$ and $\frac{S_{yy}}{E} = 0.1$, (i.e. both tensile), a_s and b_s both reduced. However, when $\frac{S_{xx}}{E} = -0.15$ and $\frac{S_{yy}}{E} = -0.1$, (i.e. both compressive), three of the six models provided us no solutions. And out of the other three, one offered a different solution than the other two.

Thus, the model you choose can influence your results significantly under certain circumstances.

In section 11.3, formulas for uniaxial loads are compared. We conclude that i) when $S_{xx} = 0$, b_e depends on b_s but not a_s and ii) when $S_{yy} = 0$, a_e depends on a_s but not b_s (i.e. for uniaxial loads, it does not matter whether the initial shape is a circle, ellipse, or a long narrow crack).

In section 11.4, displacements of points on the contour of the elliptical hole are compared. Four cases (corresponding to four loading conditions) are presented. They are 1) contraction along major axis and expansion along minor axis, 2) expansion along major axis and contraction along minor axis, 3) contraction along major axis and contraction along minor axis, and 4) expansion along major axis and expansion along minor axis. For each case, three graphs are presented. They are 1) displacement of the contour in the first quadrant, 2) displacement of selected points near the tip in the first quadrant, and 3) displacement of selected points near the top in the first quadrant.

It is noted that 1) when $S_{xx} = 0$ and $S_{yy} = 0.1E$, all six models show contraction along major axis and expansion along minor axis, 2) when $S_{xx} = 0.1E$ and $S_{yy} = 0$, all six models show expansion along major axis and contraction along minor axis, 3) when $S_{xx} = 0.15E$ and $S_{yy} = 0.1E$, all six models show contraction along major axis and contraction along minor axis. However, when $S_{xx} = -0.15E$ and $S_{yy} = -0.1E$, only model with engineering strain and model with Green strain show expansion along major axis and expansion along minor axis.

In section 11.5, zero stress points are compared. When $S_{xx} = 0$ and $S_{yy} = 0.1E$, as shown in the second column in Table 11.5.2, β_{zsp} is the lowest for Almansi strain with Kirchoff stress and the highest for Green strain with Cauchy stress. However, variation in β_{zsp} values in Table 11.5.3 where $S_{xx} = 0.1E$ and $S_{yy} = 0$ is not as much as in Table 11.5.2. Consequently, there is not much variation in magnitudes of corresponding displacements too. The last column in Table 11.5.4 where $S_{xx} = 0.15E$ and $S_{yy} = 0.1E$ indicates that there is tensile stress at the tip as well as at the top for all models. Hence, there is no zero stress point. When $S_{xx} = -0.15E$ and $S_{yy} = -0.1E$, as discussed earlier, three models did not yield any solutions for a_e and b_e. Consequently, we cannot use these three models to find the zero stress points. The model with Almansi strain and Cauchy stress shows tensile σ_{yy} at the tip (Table 11.5.5) and hence it is not an acceptable solution. The other two models show compressive stress at the tip as well as at the top. Hence, there is no zero stress point.

In section 11.6, tip stresses are compared when $S_{xx} = 0$. However, instead of keeping S_{yy} at a fixed level, we vary it from 0 to $0.5E$. Since $S_{xx} = 0$, we can use simplified formulas described in Section 11.3. Thus, b_e is calculated first and then a_e. For Singh et al model, their equation 33 and equation 34 mentioned on page 485 of their paper are used.

Finally, we calculate tip stress σ_{yy}/E at various S_{yy}/E values. These tip stress values are plotted in Figure 11.6.1. This plot in colour is also shown on the front cover of this book. From Figure 11.6.1, we observe that the Singh et al model yields the lowest tip stress. As discussed earlier, the Singh, Glinka, and Dubey model can be used only for materials with $\nu = 1$. Hence, it is discarded from further comparison.

Among the other six models, Model 2 with Green strain and Cauchy stress definitions shows the lowest tip stress

(1.64E when the far field stress is 0.5E) whereas Models 6 with Almansi strain and Kirchoff stress definitions shows the highest tip stress (5.17E when the far field stress is 0.5E). More details are discussed in section 11.6.

It should be noted that the differences in tip stress values shown above are strictly due to the definitions of strain and stress we employ. The stress-strain relations have no influence here since, in this book, we are considering only linear stress-strain relations where E is constant. Furthermore, far field stress S_{yy} is a fraction of E.

In section 11.7, limitations of models are discussed. The first limitation is the self imposed limitation that E is constant. Since in this book, the focus was on understanding non-linearity in strain-displacement relations exclusively, non-linearity in stress-strain relations was not considered. Consequently, all models developed here are applicable to only linear portions of stress-strain relations irrespective of overall material behaviour. In this book, we considered far-field stresses and induced stresses as some fraction (or multiple) of E.

The second limitation is due to the nature of mathematical relations we employed in defining strain. The incompatibility of these relations with loading conditions and/or initial geometry can be a limitation. Details and examples are presented in section 11.7.

APPENDICES

Appendix A

Expressions for $\dfrac{\partial u_x}{\partial x_s}$ and $\dfrac{\partial x_e}{\partial x_s}$

From equation (5.1.2) we know,

$$u_x = (c_e \cosh\alpha_e \cos\beta - c_e \cosh\alpha_s \cos\beta) +$$
$$(c_e \cosh\alpha_s \cos\beta - c_s \cosh\alpha_s \cos\beta)$$

Hence, $\dfrac{\partial u_x}{\partial x_s}$

$$= \frac{\partial(c_e \cosh\alpha_e \cos\beta - c_e \cosh\alpha_s \cos\beta)}{\partial x_s}$$

$$+ \frac{\partial(c_e \cosh\alpha_s \cos\beta - c_s \cosh\alpha_s \cos\beta)}{\partial x_s}$$

$$= c_e \cos\beta \, \frac{\partial(\cosh\alpha_e - \cosh\alpha_s)}{\partial x_s} + \cosh\alpha_s \cos\beta \frac{\partial(c_e - c_s)}{\partial x_s}$$

189

$$= c_e \cos\beta \left[\frac{\partial(\cosh\alpha_e)}{\partial x_s} - \frac{\partial(\cosh\alpha_s)}{\partial x_s}\right] + \cosh\alpha_s \cos\beta\left[\frac{\partial(c_e)}{\partial x_s} - \frac{\partial(c_s)}{\partial x_s}\right]$$

$$= c_e \cos\beta \left[\frac{\partial(\cosh\alpha_e)}{\partial\alpha_e}\frac{\partial(\alpha_e)}{\partial x_s}\right] - c_e \cos\beta \left[\frac{\partial(\cosh\alpha_s)}{\partial\alpha_s}\frac{\partial(\alpha_s)}{\partial x_s}\right]$$

$$+ \cosh\alpha_s \cos\beta\left[\frac{\partial(c_e)}{\partial x_e}\frac{\partial(x_e)}{\partial x_s}\right] - \cosh\alpha_s \cos\beta\left[\frac{\partial(c_s)}{\partial x_s}\right]$$

$$= c_e \cos\beta \left[(\sinh\alpha_e)\frac{\partial(\alpha_e)}{\partial x_s}\right] - c_e \cos\beta \left[(\sinh\alpha_s)\frac{\partial(\alpha_s)}{\partial x_s}\right]$$

$$+ \cosh\alpha_s \cos\beta\left[\frac{\partial(c_e)}{\partial x_e}\frac{\partial(x_e)}{\partial x_s}\right] - \cosh\alpha_s \cos\beta\left[\frac{\partial(c_s)}{\partial x_s}\right]$$

Thus,

$$\frac{\partial u_x}{\partial x_s} = \frac{\partial x_e}{\partial x_s} - 1 \qquad (5.1.1)$$

$$= c_e \cos\beta \sinh\alpha_e \left[\frac{\partial(\alpha_e)}{\partial x_e}\frac{\partial(x_e)}{\partial x_s}\right] - c_e \cos\beta \sinh\alpha_s \left[\frac{\partial(\alpha_s)}{\partial x_s}\right]$$

$$+ \cosh\alpha_s \cos\beta\left[\frac{\partial(c_e)}{\partial x_e}\frac{\partial(x_e)}{\partial x_s}\right] - \cosh\alpha_s \cos\beta\left[\frac{\partial(c_s)}{\partial x_s}\right]$$

Or, $c_e \cos\beta \sinh\alpha_s \left[\frac{\partial(\alpha_s)}{\partial x_s}\right] + \cosh\alpha_s \cos\beta\left[\frac{\partial(c_s)}{\partial x_s}\right] - 1$

$$= c_e \cos\beta\sinh\alpha_e\left[\frac{\partial(\alpha_e)}{\partial x_e}\frac{\partial(x_e)}{\partial x_s}\right] + \cosh\alpha_s \cos\beta\left[\frac{\partial(c_e)}{\partial x_e}\frac{\partial(x_e)}{\partial x_s}\right] - \frac{\partial x_e}{\partial x_s}$$

Or, $\frac{\partial x_e}{\partial x_s}\left[c_e \cos\beta \sinh\alpha_e \frac{\partial(\alpha_e)}{\partial x_e} + \cosh\alpha_s \cos\beta\frac{\partial(c_e)}{\partial x_e} - 1\right]$

$$= c_e \cos\beta \sinh\alpha_s \left[\frac{\partial(\alpha_s)}{\partial x_s}\right] + \cosh\alpha_s \cos\beta\left[\frac{\partial(c_s)}{\partial x_s}\right] - 1$$

Therefore,

$$\frac{\partial x_e}{\partial x_s} = \frac{c_e \cos\beta \sinh\alpha_s \left[\frac{\partial(\alpha_s)}{\partial x_s}\right] + \cosh\alpha_s \cos\beta\left[\frac{\partial(c_s)}{\partial x_s}\right] - 1}{c_e \cos\beta \sinh\alpha_e \left[\frac{\partial(\alpha_e)}{\partial x_e}\right] + \cosh\alpha_s \cos\beta\left[\frac{\partial(c_e)}{\partial x_e}\right] - 1}$$

$$(5.1.3)$$

Using the above expression for $\frac{\partial x_e}{\partial x_s}$, we determine

$\frac{\partial u_x}{\partial x_s}$ and $\frac{\partial x_e}{\partial x_s}$ at the tip and the top.

At the tip, $\beta = 0, x_s = a_s, y_s = 0, x_e = a_e$, and $y_e = 0$.

Hence, $f_2(x_s, y_s, c_s) = x_s^2 + y_s^2 - c_s^2 = a_s^2 - c_s^2 = b_s^2$,

$$f_1(x_s, y_s, c_s) = [f_2(x_s, y_s, c_s)]^2 + 4c_s^2\, y_s^2 = [b_s^2]^2,$$

$$f_3(x_s, y_s, c_s) = x_s^2 + y_s^2 + c_s^2 = a_s^2 + c_s^2, \text{ and}$$

$$\sqrt{f_1(x_s, y_s, c_s)}\,[f_3(x_s, y_s, c_s) + \sqrt{f_1(x_s, y_s, c_s)}]$$

$$= b_s^2[a_s^2 + c_s^2 + b_s^2] = b_s^2[a_s^2 + (a_s^2 - b_s^2) + b_s^2]$$

$$= b_s^2[2\,a_s^2] = 2\,a_s^2\, b_s^2.$$

Similarly, $f_2(x_e, y_e, c_e) = b_e^2$, $f_1(x_e, y_e, c_e) = [b_e^2]^2$,

$$f_3(x_e, y_e, c_e) = a_e^2 + c_e^2,$$

$$\sqrt{f_1(x_e, y_e, c_e)}\,[f_3(x_e, y_e, c_e) + \sqrt{f_1(x_e, y_e, c_e)}]$$

$$= 2\,a_e^2\, b_e^2, \text{ and } \cos\beta = 1.$$

Consequently, $\dfrac{\partial(\alpha_s)}{\partial x_s}$.

$$= \frac{x_s\, c_s^2 \sinh(2\alpha_s)}{\sqrt{f_1(x_s, y_s, c_s)}\,[f_3(x_s, y_s, c_s) + \sqrt{f_1(x_s, y_s, c_s)}]}$$

$$= \frac{a_s\, c_s^2 \sinh(2\alpha_s)}{2\,a_s^2\, b_s^2} = \frac{c_s^2\, 2\sinh(\alpha_s)\cosh(\alpha_s)}{2\,a_s\, b_s^2}$$

$$= \frac{c_s \sinh(\alpha_s)\, c_s \cosh(\alpha_s)}{a_s\, b_s^2} = \frac{b_s\, a_s}{a_s\, b_s^2} = \frac{1}{b_s}.$$

and $\dfrac{\partial c_s}{\partial x_s} = \dfrac{x_s \tanh\alpha_s}{\sqrt{x_s^2 \sinh^2\alpha_s + y_s^2 \cosh^2\alpha_s}} = \dfrac{a_s \tanh\alpha_s}{a_s \sinh\alpha_s}$

$$= \operatorname{sech}\alpha_s$$

Similarly, $\dfrac{\partial(\alpha_e)}{\partial x_e} = \dfrac{1}{b_e}$ and $\dfrac{\partial c_e}{\partial x_e} = \operatorname{sech}\alpha_e.$

Now we substitute these values in $\dfrac{\partial x_e}{\partial x_s}$

$$= \frac{c_e \cos\beta \, \sinh\alpha_s \, [\frac{\partial(\alpha_s)}{\partial x_s}] + \cosh\alpha_s \, \cos\beta [\frac{\partial(c_s)}{\partial x_s}] - 1}{c_e \cos\beta \, \sinh\alpha_e \, [\frac{\partial(\alpha_e)}{\partial x_e}] + \cosh\alpha_s \, \cos\beta [\frac{\partial(c_e)}{\partial x_e}] - 1}$$

$$= \frac{c_e \, (1)\sinh\alpha_s \, [\frac{1}{b_s}] + \cosh\alpha_s \, (1)[\operatorname{sech}\alpha_s] - 1}{c_e \, (1)\sinh\alpha_e \, [\frac{1}{b_e}] + \cosh\alpha_s \, (1)[\operatorname{sech}\alpha_e] - 1}$$

$$= \frac{c_e \, (1)\sinh\alpha_s \, [\frac{1}{b_s}] + 1 - 1}{1 + \cosh\alpha_s \, [\operatorname{sech}\alpha_e] - 1} = \frac{c_e \, \sinh\alpha_s \, [\frac{1}{b_s}]}{\cosh\alpha_s \, [\operatorname{sech}\alpha_e]}$$

$$= \frac{c_e \, \sinh\alpha_s \, \cosh\alpha_e}{c_s \, \sinh\alpha_s \, \cosh\alpha_s} = \frac{c_e \, \cosh\alpha_e}{c_s \, \cosh\alpha_s} = \frac{a_e}{a_s}.$$

Thus, at the tip,

$$\frac{\partial x_e}{\partial x_s} = \frac{a_e}{a_s} \text{ and } \frac{\partial u_x}{\partial x_s} = \frac{a_e - a_s}{a_s}. \qquad (5.1.4)$$

At the top, $\beta = \dfrac{\pi}{2}$ and $\cos\beta = 0.$

Hence, $\dfrac{\partial x_e}{\partial x_s}$ simplifies to $\dfrac{0 + 0 - 1}{0 + 0 - 1} = 1.$

Thus, at the top,

$$\frac{\partial x_e}{\partial x_s} = 1 \text{ and } \frac{\partial u_x}{\partial x_s} = 0. \qquad (5.1.5)$$

Appendix B

Expressions for $\dfrac{\partial u_x}{\partial y_s}$ and $\dfrac{\partial x_e}{\partial y_s}$

From equation (5.1.2) we know,

$$u_x = (c_e \cosh\alpha_e \cos\beta - c_e \cosh\alpha_s \cos\beta)+$$

$$(c_e \cosh\alpha_s \cos\beta - c_s \cosh\alpha_s \cos\beta)$$

Hence, $\dfrac{\partial u_x}{\partial y_s}$

$$= \dfrac{\partial(c_e \cosh\alpha_e \cos\beta - c_e \cosh\alpha_s \cos\beta)}{\partial y_s}$$

$$+ \dfrac{\partial(c_e \cosh\alpha_s \cos\beta - c_s \cosh\alpha_s \cos\beta)}{\partial y_s}$$

$$= c_e \cos\beta \, \dfrac{\partial(\cosh\alpha_e - \cosh\alpha_s)}{\partial y_s} + \cosh\alpha_s \cos\beta \dfrac{\partial(c_e - c_s)}{\partial y_s}$$

$$= c_e \cos\beta \left[\frac{\partial(\cosh\alpha_e)}{\partial y_s} - \frac{\partial(\cosh\alpha_s)}{\partial y_s}\right] + \cosh\alpha_s \cos\beta\left[\frac{\partial(c_e)}{\partial y_s} - \frac{\partial(c_s)}{\partial y_s}\right]$$

$$= c_e \cos\beta \left[\frac{\partial(\cosh\alpha_e)}{\partial \alpha_e}\frac{\partial(\alpha_e)}{\partial y_s}\right] - c_e \cos\beta \left[\frac{\partial(\cosh\alpha_s)}{\partial \alpha_s}\frac{\partial(\alpha_s)}{\partial y_s}\right]$$

$$+ \cosh\alpha_s \cos\beta\left[\frac{\partial(c_e)}{\partial x_e}\frac{\partial(x_e)}{\partial y_s}\right] - \cosh\alpha_s \cos\beta\left[\frac{\partial(c_s)}{\partial y_s}\right]$$

$$= c_e \cos\beta \left[(\sinh\alpha_e)\frac{\partial(\alpha_e)}{\partial y_s}\right] - c_e \cos\beta \left[(\sinh\alpha_s)\frac{\partial(\alpha_s)}{\partial y_s}\right]$$

$$+ \cosh\alpha_s \cos\beta\left[\frac{\partial(c_e)}{\partial x_e}\frac{\partial(x_e)}{\partial y_s}\right] - \cosh\alpha_s \cos\beta\left[\frac{\partial(c_s)}{\partial y_s}\right]$$

Thus,

$$\frac{\partial u_x}{\partial y_s} = \frac{\partial x_e}{\partial y_s} - \frac{\partial x_s}{\partial y_s} = \frac{\partial x_e}{\partial y_s} - 0 = \frac{\partial x_e}{\partial y_s} \qquad (5.2.1)$$

$$= c_e \cos\beta \sinh\alpha_e \left[\frac{\partial(\alpha_e)}{\partial x_e}\frac{\partial(x_e)}{\partial y_s}\right] - c_e \cos\beta \sinh\alpha_s \left[\frac{\partial(\alpha_s)}{\partial y_s}\right]$$

$$+ \cosh\alpha_s \cos\beta\left[\frac{\partial(c_e)}{\partial x_e}\frac{\partial(x_e)}{\partial y_s}\right] - \cosh\alpha_s \cos\beta\left[\frac{\partial(c_s)}{\partial y_s}\right]$$

Or, $c_e \cos\beta \sinh\alpha_s \left[\frac{\partial(\alpha_s)}{\partial y_s}\right] + \cosh\alpha_s \cos\beta\left[\frac{\partial(c_s)}{\partial y_s}\right]$

$$= c_e \cos\beta\sinh\alpha_e\left[\frac{\partial(\alpha_e)}{\partial x_e}\frac{\partial(x_e)}{\partial y_s}\right] + \cosh\alpha_s \cos\beta\left[\frac{\partial(c_e)}{\partial x_e}\frac{\partial(x_e)}{\partial y_s}\right] - \frac{\partial x_e}{\partial y_s}$$

Or, $\frac{\partial x_e}{\partial y_s}\left[c_e \cos\beta \sinh\alpha_e \left[\frac{\partial(\alpha_e)}{\partial x_e}\right] + \cosh\alpha_s \cos\beta\left[\frac{\partial(c_e)}{\partial x_e}\right] - 1\right]$

$$= c_e \cos\beta \sinh\alpha_s \left[\frac{\partial(\alpha_s)}{\partial y_s}\right] + \cosh\alpha_s \cos\beta\left[\frac{\partial(c_s)}{\partial y_s}\right]$$

Therefore,

$$\frac{\partial x_e}{\partial y_s} = \frac{c_e \cos\beta \sinh\alpha_s \left[\frac{\partial(\alpha_s)}{\partial y_s}\right] + \cosh\alpha_s \cos\beta\left[\frac{\partial(c_s)}{\partial y_s}\right]}{c_e \cos\beta \sinh\alpha_e \left[\frac{\partial(\alpha_e)}{\partial x_e}\right] + \cosh\alpha_s \cos\beta\left[\frac{\partial(c_e)}{\partial x_e}\right] - 1}$$
$$(5.2.2)$$

Using the above expression for $\dfrac{\partial x_e}{\partial y_s}$, we determine

$\dfrac{\partial u_x}{\partial y_s}$ and $\dfrac{\partial x_e}{\partial y_s}$ at the tip and the top.

At the tip, $\beta = 0, x_s = a_s, y_s = 0, x_e = a_e$, and $y_e = 0$.

Consequently, $\dfrac{\partial(\alpha_s)}{\partial y_s}$

$$= \frac{y_s\, c_s^2 \sinh(2\alpha_s)}{\sqrt{f_1(x_s, y_s, c_s)}[f_2(x_s, y_s, c_s) + \sqrt{f_1(x_s, y_s, c_s)}]} = 0$$

and $\dfrac{\partial c_s}{\partial y_s} = \dfrac{y_s \coth\alpha_s}{\sqrt{x_s^2 \sinh^2\alpha_s + y_s^2 \cosh^2\alpha_s}} = 0$ since $y_s = 0$.

From Appendix A,

we know that $\dfrac{\partial(\alpha_e)}{\partial x_e} = \dfrac{1}{b_e}$ and $\dfrac{\partial c_e}{\partial x_e} = \operatorname{sech}\alpha_e$.

Now we substitute these values in $\frac{\partial x_e}{\partial y_s}$

$$= \frac{c_e \cos\beta \sinh\alpha_s \left[\frac{\partial(\alpha_s)}{\partial y_s}\right] + \cosh\alpha_s \cos\beta\left[\frac{\partial(c_s)}{\partial y_s}\right]}{c_e \cos\beta \sinh\alpha_e \left[\frac{\partial(\alpha_e)}{\partial x_e}\right] + \cosh\alpha_s \cos\beta\left[\frac{\partial(c_e)}{\partial x_e}\right] - 1}$$

$$= \frac{c_e \cos\beta \sinh\alpha_s\, [0] + \cosh\alpha_s \cos\beta\, [0]}{c_e \cos\beta \sinh\alpha_e\, [\frac{1}{b_e}] + \cosh\alpha_s \cos\beta[\operatorname{sech}\alpha_e] - 1} = 0$$

Thus, at the tip,

$$\frac{\partial x_e}{\partial y_s} = 0 \text{ and } \frac{\partial u_x}{\partial y_s} = \frac{\partial x_e}{\partial y_s} = 0. \qquad (5.2.3)$$

At the top, $\beta = \frac{\pi}{2}$ and $\cos\beta = 0$.

Hence, $\frac{\partial x_e}{\partial y_s}$ simplifies to $\frac{0+0}{0+0-1} = 0$.

Thus, at the top,

$$\frac{\partial x_e}{\partial y_s} = 0 \text{ and } \frac{\partial u_x}{\partial y_s} = \frac{\partial x_e}{\partial y_s} = 0. \qquad (5.2.4)$$

Appendix C

Expressions for $\dfrac{\partial u_y}{\partial x_s}$ and $\dfrac{\partial y_e}{\partial x_s}$

From equation (5.3.2) we know,

$$u_y = (c_e \sinh\alpha_e \sin\beta - c_e \sinh\alpha_s \sin\beta) +$$
$$(c_e \sinh\alpha_s \sin\beta - c_s \sinh\alpha_s \sin\beta).$$

Hence, $\dfrac{\partial u_y}{\partial x_s}$

$$= \frac{\partial(c_e \sinh\alpha_e \sin\beta - c_e \sinh\alpha_s \sin\beta)}{\partial x_s}$$

$$+ \frac{\partial(c_e \sinh\alpha_s \sin\beta - c_s \sinh\alpha_s \sin\beta)}{\partial x_s}$$

$$= c_e \sin\beta \, \frac{\partial(\sinh\alpha_e - \sinh\alpha_s)}{\partial x_s} + \sinh\alpha_s \sin\beta \, \frac{\partial(c_e - c_s)}{\partial x_s}$$

197

$$= c_e \sin\beta \left[\frac{\partial(\sinh\alpha_e)}{\partial x_s} - \frac{\partial(\sinh\alpha_s)}{\partial x_s}\right] + \sinh\alpha_s \sin\beta\left[\frac{\partial(c_e)}{\partial x_s} - \frac{\partial(c_s)}{\partial x_s}\right]$$

$$= c_e \sin\beta \left[\frac{\partial(\sinh\alpha_e)}{\partial x_s}\right] - c_e \sin\beta \left[\frac{\partial(\sinh\alpha_s)}{\partial x_s}\right]$$

$$+ \sinh\alpha_s \sin\beta\left[\frac{\partial(c_e)}{\partial x_s}\right] - \sinh\alpha_s \sin\beta\left[\frac{\partial(c_s)}{\partial x_s}\right]$$

$$= c_e \sin\beta \left[\frac{\partial(\sinh\alpha_e)}{\partial(\alpha_e)}\frac{\partial(\alpha_e)}{\partial x_s}\right] - c_e \sin\beta \left[\frac{\partial(\sinh\alpha_s)}{\partial(\alpha_s)}\frac{\partial(\alpha_s)}{\partial x_s}\right]$$

$$+ \sinh\alpha_s \sin\beta\left[\frac{\partial(c_e)}{\partial y_e}\frac{\partial(y_e)}{\partial x_s}\right] - \sinh\alpha_s \sin\beta\left[\frac{\partial(c_s)}{\partial x_s}\right]$$

$$= c_e \sin\beta \left[(\cosh\alpha_e)\frac{\partial(\alpha_e)}{\partial x_s}\right] - c_e \sin\beta \left[(\cosh\alpha_s)\frac{\partial(\alpha_s)}{\partial x_s}\right]$$

$$+ \sinh\alpha_s \sin\beta\left[\frac{\partial(c_e)}{\partial y_e}\frac{\partial(y_e)}{\partial x_s}\right] - \sinh\alpha_s \sin\beta\left[\frac{\partial(c_s)}{\partial x_s}\right]$$

Thus,

$$\frac{\partial u_y}{\partial x_s} = \frac{\partial y_e}{\partial x_s} - \frac{\partial y_s}{\partial x_s} = \frac{\partial y_e}{\partial x_s} - 0 = \frac{\partial y_e}{\partial x_s} \qquad (5.3.1)$$

$$= c_e \sin\beta \, (\cosh\alpha_e) \left[\frac{\partial(\alpha_e)}{\partial y_e}\frac{\partial(y_e)}{\partial x_s}\right] - c_e \sin\beta \left[(\cosh\alpha_s)\frac{\partial(\alpha_s)}{\partial x_s}\right]$$

$$+ \sinh\alpha_s \sin\beta\left[\frac{\partial(c_e)}{\partial y_e}\frac{\partial(y_e)}{\partial x_s}\right] - \sinh\alpha_s \sin\beta\left[\frac{\partial(c_s)}{\partial x_s}\right]$$

Or, $c_e \sin\beta \cosh\alpha_s \left[\frac{\partial(\alpha_s)}{\partial x_s}\right] + \sinh\alpha_s \sin\beta\left[\frac{\partial(c_s)}{\partial x_s}\right]$

$$= c_e \sin\beta \cosh\alpha_e \left[\frac{\partial(\alpha_e)}{\partial y_e}\frac{\partial(y_e)}{\partial x_s}\right] + \sinh\alpha_s \sin\beta\left[\frac{\partial(c_e)}{\partial y_e}\frac{\partial(y_e)}{\partial x_s}\right] - \frac{\partial y_e}{\partial x_s}$$

Or, $\frac{\partial y_e}{\partial x_s}\left[c_e \sin\beta \cosh\alpha_e \left[\frac{\partial(\alpha_e)}{\partial y_e}\right] + \sinh\alpha_s \sin\beta\left[\frac{\partial(c_e)}{\partial y_e}\right] - 1\right]$

$$= c_e \sin\beta \cosh\alpha_s \left[\frac{\partial(\alpha_s)}{\partial x_s}\right] + \sinh\alpha_s \sin\beta\left[\frac{\partial(c_s)}{\partial x_s}\right]$$

Therefore,

$$\frac{\partial y_e}{\partial x_s} = \frac{c_e \sin\beta \cosh\alpha_s \left[\frac{\partial(\alpha_s)}{\partial x_s}\right] + \sinh\alpha_s \sin\beta\left[\frac{\partial(c_s)}{\partial x_s}\right]}{\left[c_e \sin\beta \cosh\alpha_e \left[\frac{\partial(\alpha_e)}{\partial y_e}\right] + \sinh\alpha_s \sin\beta\left[\frac{\partial(c_e)}{\partial y_e}\right] - 1\right]}$$

$$(5.3.3)$$

Using the above expression for $\dfrac{\partial y_e}{\partial x_s}$, we determine

$\dfrac{\partial u_y}{\partial x_s}$ and $\dfrac{\partial y_e}{\partial x_s}$ at the tip and the top.

At the tip, $\beta = 0$ and $\sin\beta = 0$.

Hence, $\dfrac{\partial y_e}{\partial x_s}$ simplifies to $\dfrac{0+0}{0+0-1} = 0$.

Thus, at the tip,

$$\frac{\partial y_e}{\partial x_s} = 0 \text{ and } \frac{\partial u_y}{\partial x_s} = \frac{\partial y_e}{\partial x_s} = 0. \qquad (5.3.4)$$

At the top, $x_e = 0$, and $y_e = b_e$

Hence,

$$f_2(x_e, y_e, c_e) = x_e^2 + y_e^2 - c_e^2 = 0 + b_e^2 - c_e^2,$$

$$f_1(x_e, y_e, c_e) = [f_2(x_e, y_e, c_e)]^2 + 4c_e^2 \, y_e^2$$

$$= [b_e^2 - c_e^2]^2 + 4c_e^2 \, b_e^2 = [b_e^2 + c_e^2]^2 = [a_e^2]^2, \text{ and}$$

$$\sqrt{f_1(x_e, y_e, c_e)}[f_2(x_e, y_e, c_e) + \sqrt{f_1(x_e, y_e, c_e)}]$$

$$= [b_e^2 + c_e^2][b_e^2 - c_e^2 + b_e^2 + c_e^2] = 2 \, a_e^2 \, b_e^2.$$

Consequently, $\dfrac{\partial \alpha_e}{\partial y_e}$

$$= \frac{y_e \, c_e^2 \sinh(2\alpha_e)}{\sqrt{f_1(x_e, y_e, c_e)}(f_2(x_e, y_e, c_e) + \sqrt{f_1(x_e, y_e, c_e)})}$$

$$= \frac{b_e \, c_e^2 \, 2 \sinh(\alpha_e) \cosh(\alpha_e)}{2 \, a_e^2 \, b_e^2}$$

$$= \frac{c_e \sinh(\alpha_e)\, c_e \cosh(\alpha_e)}{a_e^2\, b_e} = \frac{b_e\, a_e}{a_e^2\, b_e} = \frac{1}{a_e}$$

And, $\dfrac{\partial c_e}{\partial y_e} = \dfrac{y_e \coth\alpha_e}{\sqrt{x_e^2 \sinh^2\alpha_e + y_e^2 \cosh^2\alpha_e}} = \dfrac{y_e \coth\alpha_e}{\sqrt{0 + y_e^2 \cosh^2\alpha_e}}$

$$= \frac{y_e \coth\alpha_e}{y_e \cosh\alpha_e} = \frac{1}{\sinh\alpha_e} = \frac{c_e}{b_e}$$

Also, at the top, $x_s = 0$.

Hence, $\dfrac{\partial(\alpha_s)}{\partial x_s}$

$$= \frac{x_s\, c_s^2 \sinh(2\alpha_s)}{\sqrt{f_1(x_s, y_s, c_s)}[f_3(x_s, y_s, c_s) + \sqrt{f_1(x_s, y_s, c_s)}]} = 0.$$

And, $\dfrac{\partial(c_s)}{\partial x_s} = \dfrac{x_s \tanh\alpha_s}{\sqrt{x_s^2 \sinh^2\alpha_s + y_s^2 \cosh^2\alpha_s}} = 0.$

Substituting these values in equation (5.3.3), we obtain

$$\frac{\partial y_e}{\partial x_s} = \frac{c_e \sin\beta \cosh\alpha_s \left[\frac{\partial(\alpha_s)}{\partial x_s}\right] + \sinh\alpha_s \sin\beta \left[\frac{\partial(c_s)}{\partial x_s}\right]}{[c_e \sin\beta \cosh\alpha_e \left[\frac{\partial(\alpha_e)}{\partial y_e}\right] + \sinh\alpha_s \sin\beta \left[\frac{\partial(c_e)}{\partial y_e}\right] - 1]}$$

$$= \frac{c_e \sin\beta \cosh\alpha_s\, [0] + \sinh\alpha_s \sin\beta [0]}{[c_e \sin\beta \cosh\alpha_e\, [\frac{1}{a_e}] + \sinh\alpha_s \sin\beta [\frac{c_e}{b_e}] - 1]}$$

$$= \frac{[0] + [0]}{[1] + \sinh\alpha_s\, (1)[\frac{c_e}{b_e}] - 1]} = 0.$$

Thus, at the top,

$$\frac{\partial y_e}{\partial x_s} = 0 \text{ and } \frac{\partial u_y}{\partial x_s} = \frac{\partial y_e}{\partial x_s} = 0. \qquad (5.3.5)$$

Appendix D

Expressions for $\dfrac{\partial u_y}{\partial y_s}$ and $\dfrac{\partial y_e}{\partial y_s}$

From equation (5.3.2) we know,

$$u_y = (c_e \sinh\alpha_e \sin\beta - c_e \sinh\alpha_s \sin\beta)+$$

$$(c_e \sinh\alpha_s \sin\beta - c_s \sinh\alpha_s \sin\beta).$$

Hence, $\dfrac{\partial u_y}{\partial y_s}$

$$= \dfrac{\partial(c_e \sinh\alpha_e \sin\beta - c_e \sinh\alpha_s \sin\beta)}{\partial y_s}$$

$$+ \dfrac{\partial(c_e \sinh\alpha_s \sin\beta - c_s \sinh\alpha_s \sin\beta)}{\partial y_s}$$

$$= c_e \sin\beta \, \dfrac{\partial(\sinh\alpha_e - \sinh\alpha_s)}{\partial y_s} + \sinh\alpha_s \sin\beta \, \dfrac{\partial(c_e - c_s)}{\partial y_s}$$

$$= c_e \sin\beta \left[\frac{\partial(\sinh\alpha_e)}{\partial y_s} - \frac{\partial(\sinh\alpha_s)}{\partial y_s}\right] + \sinh\alpha_s \sin\beta\left[\frac{\partial(c_e)}{\partial y_s} - \frac{\partial(c_s)}{\partial y_s}\right]$$

$$= c_e \sin\beta \left[\frac{\partial(\sinh\alpha_e)}{\partial y_s}\right] - c_e \sin\beta \left[\frac{\partial(\sinh\alpha_s)}{\partial y_s}\right]$$

$$+ \sinh\alpha_s \sin\beta\left[\frac{\partial(c_e)}{\partial y_s}\right] - \sinh\alpha_s \sin\beta\left[\frac{\partial(c_s)}{\partial y_s}\right]$$

$$= c_e \sin\beta \left[\frac{\partial(\sinh\alpha_e)}{\partial(\alpha_e)}\frac{\partial(\alpha_e)}{\partial y_s}\right] - c_e \sin\beta \left[\frac{\partial(\sinh\alpha_s)}{\partial(\alpha_s)}\frac{\partial(\alpha_s)}{\partial y_s}\right]$$

$$+ \sinh\alpha_s \sin\beta\left[\frac{\partial(c_e)}{\partial y_e}\frac{\partial(y_e)}{\partial y_s}\right] - \sinh\alpha_s \sin\beta\left[\frac{\partial(c_s)}{\partial y_s}\right]$$

$$= c_e \sin\beta \left[(\cosh\alpha_e)\frac{\partial(\alpha_e)}{\partial y_s}\right] - c_e \sin\beta \left[(\cosh\alpha_s)\frac{\partial(\alpha_s)}{\partial y_s}\right]$$

$$+ \sinh\alpha_s \sin\beta\left[\frac{\partial(c_e)}{\partial y_e}\frac{\partial(y_e)}{\partial y_s}\right] - \sinh\alpha_s \sin\beta\left[\frac{\partial(c_s)}{\partial y_s}\right]$$

Thus,

$$\frac{\partial u_y}{\partial y_s} = \frac{\partial y_e}{\partial y_s} - \frac{\partial y_s}{\partial y_s} = \frac{\partial y_e}{\partial y_s} - 1 \qquad (5.4.1)$$

$$= c_e \sin\beta \,(\cosh\alpha_e) \left[\frac{\partial(\alpha_e)}{\partial y_e}\frac{\partial(y_e)}{\partial y_s}\right] - c_e \sin\beta \left[(\cosh\alpha_s)\frac{\partial(\alpha_s)}{\partial y_s}\right]$$

$$+ \sinh\alpha_s \sin\beta\left[\frac{\partial(c_e)}{\partial y_e}\frac{\partial(y_e)}{\partial y_s}\right] - \sinh\alpha_s \sin\beta\left[\frac{\partial(c_s)}{\partial y_s}\right]$$

Or, $c_e \sin\beta \cosh\alpha_s \left[\frac{\partial(\alpha_s)}{\partial y_s}\right] + \sinh\alpha_s \sin\beta\left[\frac{\partial(c_s)}{\partial y_s}\right] - 1$

$$= c_e \sin\beta \cosh\alpha_e \left[\frac{\partial(\alpha_e)}{\partial y_e}\frac{\partial(y_e)}{\partial y_s}\right] + \sinh\alpha_s \sin\beta\left[\frac{\partial(c_e)}{\partial y_e}\frac{\partial(y_e)}{\partial y_s}\right] - \frac{\partial y_e}{\partial y_s}$$

Or, $\frac{\partial y_e}{\partial y_s}\left[c_e \sin\beta \cosh\alpha_e \left[\frac{\partial(\alpha_e)}{\partial y_e}\right] + \sinh\alpha_s \sin\beta\left[\frac{\partial(c_e)}{\partial y_e}\right] - 1\right]$

$$= c_e \sin\beta \cosh\alpha_s \left[\frac{\partial(\alpha_s)}{\partial y_s}\right] + \sinh\alpha_s \sin\beta\left[\frac{\partial(c_s)}{\partial y_s}\right] - 1$$

Therefore,

$$\frac{\partial y_e}{\partial y_s} = \frac{c_e \sin\beta \cosh\alpha_s \left[\frac{\partial(\alpha_s)}{\partial y_s}\right] + \sinh\alpha_s \sin\beta\left[\frac{\partial(c_s)}{\partial y_s}\right] - 1}{\left[c_e \sin\beta \cosh\alpha_e \left[\frac{\partial(\alpha_e)}{\partial y_e}\right] + \sinh\alpha_s \sin\beta\left[\frac{\partial(c_e)}{\partial y_e}\right] - 1\right]}$$

$$(5.4.2)$$

Using the above expression for $\dfrac{\partial y_e}{\partial y_s}$, we determine

$\dfrac{\partial u_y}{\partial y_s}$ and $\dfrac{\partial y_e}{\partial y_s}$ at the tip and the top.

At the tip, $\beta = 0$ and $\sin\beta = 0$.

Hence, $\dfrac{\partial y_e}{\partial y_s}$ simplifies to $\dfrac{0+0-1}{0+0-1} = 1$.

Thus, at the tip,

$$\frac{\partial y_e}{\partial y_s} = 1 \text{ and } \frac{\partial u_y}{\partial y_s} = \frac{\partial y_e}{\partial y_s} - 1 = 0. \qquad (5.4.3)$$

At the top, $x_s = 0$ and $y_s = b_s$

Hence,

$$f_2(x_s, y_s, c_s) = x_s^2 + y_s^2 - c_s^2 = 0 + b_s^2 - c_s^2,$$
$$f_1(x_s, y_s, c_s) = [f_2(x_s, y_s, c_s)]^2 + 4c_s^2\, y_s^2$$
$$= [b_s^2 - c_s^2]^2 + 4c_s^2\, b_s^2 = [b_s^2 + c_s^2]^2 = [a_s^2]^2, \text{ and}$$
$$\sqrt{f_1(x_s, y_s, c_s)}[f_2(x_s, y_s, c_s) + \sqrt{f_1(x_s, y_s, c_s)}]$$
$$= [b_s^2 + c_s^2][b_s^2 - c_s^2 + b_s^2 + c_s^2] = [b_s^2 + c_s^2][2\, b_s^2]$$
$$= 2\, a_s^2\, b_s^2.$$

Consequently, $\dfrac{\partial \alpha_s}{\partial y_s}$

$$= \frac{y_s\, c_s^2 \sinh(2\alpha_s)}{\sqrt{f_1(x_s, y_s, c_s)}(f_2(x_s, y_s, c_s) + \sqrt{f_1(x_s, y_s, c_s)})}$$
$$= \frac{b_s\, c_s^2\, 2 \sinh(\alpha_s) \cosh(\alpha_s)}{2\, a_s^2\, b_s^2}$$

$$= \frac{c_s \sinh(\alpha_s) \, c_s \cosh(\alpha_s)}{a_s^2 \, b_s} = \frac{b_s \, a_s}{a_s^2 \, b_s} = \frac{1}{a_s}.$$

And, $\dfrac{\partial c_s}{\partial y_s} = \dfrac{y_s \coth\alpha_s}{\sqrt{x_s^2 \sinh^2\alpha_s + y_s^2 \cosh^2\alpha_s}} = \dfrac{y_s \coth\alpha_s}{\sqrt{0 + y_s^2 \cosh^2\alpha_s}}$

$$= \frac{y_s \coth\alpha_s}{y_s \cosh\alpha_s} = \frac{1}{\sinh\alpha_s} = \frac{c_s}{b_s}.$$

Similarly, $\dfrac{\partial \alpha_e}{\partial y_e} = \dfrac{1}{a_e}$ and $\dfrac{\partial c_e}{\partial y_e} = \dfrac{c_e}{b_e}$.

Furthermore, $\beta = \pi/2$ and $\sin\beta = 1$.

Substituting these values in equation (5.4.2), we obtain

$\dfrac{\partial y_e}{\partial y_s}$

$$= \frac{c_e \sin\beta \cosh\alpha_s \left[\frac{\partial(\alpha_s)}{\partial y_s}\right] + \sinh\alpha_s \sin\beta\left[\frac{\partial(c_s)}{\partial y_s}\right] - 1}{\left[c_e \sin\beta \cosh\alpha_e \left[\frac{\partial(\alpha_e)}{\partial y_e}\right] + \sinh\alpha_s \sin\beta\left[\frac{\partial(c_e)}{\partial y_e}\right] - 1\right]}$$

$$= \frac{c_e [1] \cosh\alpha_s \left[\frac{1}{a_s}\right] + \sinh\alpha_s [1]\left[\frac{c_s}{b_s}\right] - 1}{\left[c_e [1] \cosh\alpha_e \left[\frac{1}{a_e}\right] + \sinh\alpha_s [1]\left[\frac{c_e}{b_e}\right] - 1\right]}$$

$$= \frac{c_e \cosh\alpha_s \left[\frac{1}{a_s}\right] + b_s \left[\frac{1}{b_s}\right] - 1}{\left[a_e \left[\frac{1}{a_e}\right] + \sinh\alpha_s \left[\frac{c_e}{b_e}\right] - 1\right]} = \frac{c_e \cosh\alpha_s \left[\frac{1}{a_s}\right] + 1 - 1}{\left[1 + \sinh\alpha_s \left[\frac{c_e}{b_e}\right] - 1\right]}$$

$$= \frac{c_e \cosh\alpha_s \, b_e}{a_s \sinh\alpha_s \, c_e} = \frac{\cosh\alpha_s \, b_e}{[c_s \cosh\alpha_s] \sinh\alpha_s} = \frac{b_e}{c_s \sinh\alpha_s} = \frac{b_e}{b_s}.$$

Thus, at the top,

$$\frac{\partial y_e}{\partial y_s} = \frac{b_e}{b_s} \quad \text{and} \quad \frac{\partial u_y}{\partial y_s} = \frac{\partial y_e}{\partial y_s} - 1 = \frac{b_e - b_s}{b_s}. \qquad (5.4.4)$$

Appendix E

Engineering strain

We know from equation (8.1.2),

$\frac{b_e}{b_s} - 1 = \frac{\nu}{E}[S_{yy} - S_{xx}(1 + \frac{2b_e}{a_e})] = [\frac{\nu}{E}S_{yy} - \frac{\nu}{E}S_{xx}(1 + \frac{2b_e}{a_e})]$.

Or, $b_e = b_s + b_s[\frac{\nu}{E}S_{yy} - \frac{\nu}{E}S_{xx}(1 + \frac{2b_e}{a_e})]$

$= b_s + b_s[\frac{\nu}{E}S_{yy} - \frac{\nu}{E}S_{xx} - \frac{\nu}{E}S_{xx}\frac{2b_e}{a_e}]$.

Or, $b_e + b_s[\frac{\nu}{E}S_{xx}\frac{2b_e}{a_e}] = b_s + b_s[\frac{\nu}{E}S_{yy} - \frac{\nu}{E}S_{xx}]$.

Or, $b_e\, a_e + 2b_e b_s[\frac{\nu}{E}S_{xx}] = a_e b_s[1 + \frac{\nu}{E}S_{yy} - \frac{\nu}{E}S_{xx}]$.

Or, $b_e[a_e + 2b_s\frac{\nu}{E}S_{xx}] = a_e b_s[1 + \frac{\nu}{E}S_{yy} - \frac{\nu}{E}S_{xx}]$.

We observe that even if two surfaces are closely touching each other, b_s exists, however small it may be. We can argue that if $b_s = 0$ then there is no crack. Hence, $b_s > 0$. Furthermore, $a_e + 2b_s\frac{\nu}{E}S_{xx} = 0$ only if $S_{xx} = -\left[\frac{a_e}{2b_s}\right]\left[\frac{E}{\nu}\right]$.

Since S_{xx} cannot be in such a large magnitude, we exclude such a possibility and divide both sides by $[a_e + 2b_s \frac{\nu}{E} S_{xx}]$ so that

$$b_e = \frac{a_e b_s [1 + \frac{\nu}{E} S_{yy} - \frac{\nu}{E} S_{xx}]}{[a_e + 2b_s \frac{\nu}{E} S_{xx}]}. \qquad (8.2.1)$$

We know from equation (8.1.1),

$\frac{a_e}{a_s} - 1 = \frac{\nu}{E}[S_{xx} - S_{yy}(1 + \frac{2a_e}{b_e})] = [\frac{\nu}{E} S_{xx} - \frac{\nu}{E} S_{yy}(1 + \frac{2a_e}{b_e})].$

Or, $a_e = a_s + a_s[\frac{\nu}{E} S_{xx} - \frac{\nu}{E} S_{yy}(1 + \frac{2a_e}{b_e})]$

$\qquad = a_s + a_s[\frac{\nu}{E} S_{xx} - \frac{\nu}{E} S_{yy} - \frac{\nu}{E} S_{yy} \frac{2a_e}{b_e}].$

Or, $a_e + a_s[\frac{\nu}{E} S_{yy} 2a_e][\frac{1}{b_e}] = a_s + a_s[\frac{\nu}{E} S_{xx} - \frac{\nu}{E} S_{yy}].$

We note that S_{xx} must be less than $\frac{E}{\nu}$ since such large far field stresses are not practical.

Hence, $S_{xx} < \frac{E}{\nu} + S_{yy}$. Or, $\frac{E}{\nu} + S_{yy} - S_{xx} > 0$. Or, $1 + \frac{\nu}{E} S_{yy} - \frac{\nu}{E} S_{xx} > 0$.

Since $b_s > 0$ it implies that $b_s(1 + \frac{\nu}{E} S_{yy} - \frac{\nu}{E} S_{xx}) > 0$.

Therefore, $a_e + 2a_s[\frac{\nu}{E} S_{yy}]a_e \left[\dfrac{(a_e + 2b_s \frac{\nu}{E} S_{xx})}{a_e b_s(1 + \frac{\nu}{E} S_{yy} - \frac{\nu}{E} S_{xx})} \right]$

$\qquad = a_s + a_s[\frac{\nu}{E} S_{xx} - \frac{\nu}{E} S_{yy}]$, using equation (8.2.1).

Or, $a_e + \left[\dfrac{2a_s[\frac{\nu}{E} S_{yy}](a_e)}{b_s(1 + \frac{\nu}{E} S_{yy} - \frac{\nu}{E} S_{xx})} \right] + \left[\dfrac{2a_s[\frac{\nu}{E} S_{yy}](2b_s \frac{\nu}{E} S_{xx})}{b_s(1 + \frac{\nu}{E} S_{yy} - \frac{\nu}{E} S_{xx})} \right]$

$\qquad = a_s + a_s[\frac{\nu}{E} S_{xx} - \frac{\nu}{E} S_{yy}].$

Or, $a_e + a_e \left[\dfrac{2a_s[\frac{\nu}{E}S_{yy}]}{b_s(1 + \frac{\nu}{E}S_{yy} - \frac{\nu}{E}S_{xx})} \right]$

$$= a_s[1 + \tfrac{\nu}{E}S_{xx} - \tfrac{\nu}{E}S_{yy}] - \left[\dfrac{2a_s[\frac{\nu}{E}S_{yy}](2b_s\frac{\nu}{E}S_{xx})}{b_s(1 + \frac{\nu}{E}S_{yy} - \frac{\nu}{E}S_{xx})} \right].$$

Or, $a_e \left[\dfrac{b_s(1 + \frac{\nu}{E}S_{yy} - \frac{\nu}{E}S_{xx})}{b_s(1 + \frac{\nu}{E}S_{yy} - \frac{\nu}{E}S_{xx})} \right] + a_e \left[\dfrac{2a_s(\frac{\nu}{E}S_{yy})}{b_s(1 + \frac{\nu}{E}S_{yy} - \frac{\nu}{E}S_{xx})} \right]$

$$= a_s[1 + \tfrac{\nu}{E}S_{xx} - \tfrac{\nu}{E}S_{yy}]\left[\dfrac{b_s(1 + \frac{\nu}{E}S_{yy} - \frac{\nu}{E}S_{xx})}{b_s(1 + \frac{\nu}{E}S_{yy} - \frac{\nu}{E}S_{xx})} \right]$$

$$- \left[\dfrac{2a_s[\frac{\nu}{E}S_{yy}](2b_s\frac{\nu}{E}S_{xx})}{b_s(1 + \frac{\nu}{E}S_{yy} - \frac{\nu}{E}S_{xx})} \right].$$

Or, $a_e \left[\dfrac{b_s(1 + \frac{\nu}{E}S_{yy} - \frac{\nu}{E}S_{xx}) + 2a_s(\frac{\nu}{E}S_{yy})}{b_s(1 + \frac{\nu}{E}S_{yy} - \frac{\nu}{E}S_{xx})} \right]$

$$= a_s b_s \left[\dfrac{[1+\frac{\nu}{E}S_{xx}-\frac{\nu}{E}S_{yy}](1+\frac{\nu}{E}S_{yy}-\frac{\nu}{E}S_{xx})-2[\frac{\nu}{E}S_{yy}](2\frac{\nu}{E}S_{xx})}{b_s(1+\frac{\nu}{E}S_{yy}-\frac{\nu}{E}S_{xx})} \right].$$

As noted earlier, $b_s > 0$ and $b_s(1 + \frac{\nu}{E}S_{yy} - \frac{\nu}{E}S_{xx}) > 0$. Also, $a_s > 0$, $E > 0$, $\nu > 0$, and $\frac{\nu}{E} > 0$.

Hence, $b_s(1 + \frac{\nu}{E}S_{yy} - \frac{\nu}{E}S_{xx}) + 2a_s(\frac{\nu}{E}S_{yy}) > 0$.

We now divide both sides by

$b_s(1 + \frac{\nu}{E}S_{yy} - \frac{\nu}{E}S_{xx}) + 2a_s(\frac{\nu}{E}S_{yy})$ so that

a_e

$$= a_s b_s \left[\dfrac{[1+\frac{\nu}{E}S_{xx}-\frac{\nu}{E}S_{yy}](1+\frac{\nu}{E}S_{yy}-\frac{\nu}{E}S_{xx})-2[\frac{\nu}{E}S_{yy}](2\frac{\nu}{E}S_{xx})}{b_s(1+\frac{\nu}{E}S_{yy}-\frac{\nu}{E}S_{xx})+2a_s(\frac{\nu}{E}S_{yy})} \right]$$

$$= a_s b_s \left[\dfrac{[1+(\frac{\nu}{E}S_{xx}-\frac{\nu}{E}S_{yy})][1-(-\frac{\nu}{E}S_{yy}+\frac{\nu}{E}S_{xx})]-4[\frac{\nu}{E}S_{yy}](\frac{\nu}{E}S_{xx})}{b_s(1+\frac{\nu}{E}S_{yy}-\frac{\nu}{E}S_{xx})+2a_s(\frac{\nu}{E}S_{yy})} \right]$$

$$= a_s b_s \left[\dfrac{[1-(\frac{\nu}{E}S_{xx}-\frac{\nu}{E}S_{yy})^2]-4[\frac{\nu}{E}S_{yy}](\frac{\nu}{E}S_{xx})}{b_s(1+\frac{\nu}{E}S_{yy}-\frac{\nu}{E}S_{xx})+2a_s(\frac{\nu}{E}S_{yy})} \right]$$

$$= a_s b_s \left[\frac{[1-(\frac{\nu}{E}S_{xx})^2-(\frac{\nu}{E}S_{yy})^2+2\frac{\nu}{E}S_{xx}\frac{\nu}{E}S_{yy}]-4[\frac{\nu}{E}S_{yy}](\frac{\nu}{E}S_{xx})}{b_s(1+\frac{\nu}{E}S_{yy}-\frac{\nu}{E}S_{xx})+2a_s(\frac{\nu}{E}S_{yy})} \right]$$

$$= a_s b_s \left[\frac{[1-(\frac{\nu}{E}S_{xx})^2-(\frac{\nu}{E}S_{yy})^2-2\frac{\nu}{E}S_{xx}\frac{\nu}{E}S_{yy}]}{b_s(1+\frac{\nu}{E}S_{yy}-\frac{\nu}{E}S_{xx})+2a_s(\frac{\nu}{E}S_{yy})} \right].$$

$$= a_s b_s \left[\frac{[1-(\frac{\nu}{E}S_{xx}+\frac{\nu}{E}S_{yy})^2]}{b_s(1+\frac{\nu}{E}S_{yy}-\frac{\nu}{E}S_{xx})+2a_s(\frac{\nu}{E}S_{yy})} \right].$$

Or,

$$a_e = a_s b_s \left[\frac{(1-\frac{\nu}{E}S_{xx}-\frac{\nu}{E}S_{yy})(1+\frac{\nu}{E}S_{xx}+\frac{\nu}{E}S_{yy})}{b_s(1+\frac{\nu}{E}S_{yy}-\frac{\nu}{E}S_{xx})+2a_s(\frac{\nu}{E}S_{yy})} \right]. \quad (8.2.2)$$

Substituting this value for a_e in equation (8.2.1), we obtain

$$b_e[a_e + 2b_s\frac{\nu}{E}S_{xx}] = a_e b_s[1 + \frac{\nu}{E}S_{yy} - \frac{\nu}{E}S_{xx}].$$

Or, $b_e[a_s b_s \left[\frac{(1-\frac{\nu}{E}S_{xx}-\frac{\nu}{E}S_{yy})(1+\frac{\nu}{E}S_{xx}+\frac{\nu}{E}S_{yy})}{b_s(1+\frac{\nu}{E}S_{yy}-\frac{\nu}{E}S_{xx})+2a_s(\frac{\nu}{E}S_{yy})} \right] + 2b_s\frac{\nu}{E}S_{xx}]$

$$= a_s b_s \left[\frac{(1-\frac{\nu}{E}S_{xx}-\frac{\nu}{E}S_{yy})(1+\frac{\nu}{E}S_{xx}+\frac{\nu}{E}S_{yy})}{b_s(1+\frac{\nu}{E}S_{yy}-\frac{\nu}{E}S_{xx})+2a_s(\frac{\nu}{E}S_{yy})} \right] b_s[1 + \frac{\nu}{E}S_{yy} - \frac{\nu}{E}S_{xx}].$$

Or, $b_e[a_s \left[\frac{[1-(\frac{\nu}{E}S_{xx}+\frac{\nu}{E}S_{yy})][1+(\frac{\nu}{E}S_{xx}+\frac{\nu}{E}S_{yy})]}{b_s(1+\frac{\nu}{E}S_{yy}-\frac{\nu}{E}S_{xx})+2a_s(\frac{\nu}{E}S_{yy})} \right] + 2\frac{\nu}{E}S_{xx}]$

$$= a_s b_s \left[\frac{(1-\frac{\nu}{E}S_{xx}-\frac{\nu}{E}S_{yy})(1+\frac{\nu}{E}S_{xx}+\frac{\nu}{E}S_{yy})[1+\frac{\nu}{E}S_{yy}-\frac{\nu}{E}S_{xx}]}{b_s(1+\frac{\nu}{E}S_{yy}-\frac{\nu}{E}S_{xx})+2a_s(\frac{\nu}{E}S_{yy})} \right].$$

As above, $b_s(1 + \frac{\nu}{E}S_{yy} - \frac{\nu}{E}S_{xx}) + 2a_s(\frac{\nu}{E}S_{yy}) > 0$ so that

$$b_e a_s \left[\frac{[1-(\frac{\nu}{E}S_{xx}+\frac{\nu}{E}S_{yy})^2]}{b_s(1+\frac{\nu}{E}S_{yy}-\frac{\nu}{E}S_{xx})+2a_s(\frac{\nu}{E}S_{yy})} \right]$$

$$+ 2b_e\frac{\nu}{E}S_{xx}\frac{b_s(1+\frac{\nu}{E}S_{yy}-\frac{\nu}{E}S_{xx})+2a_s(\frac{\nu}{E}S_{yy})}{b_s(1+\frac{\nu}{E}S_{yy}-\frac{\nu}{E}S_{xx})+2a_s(\frac{\nu}{E}S_{yy})}$$

$$= a_s b_s \left[\frac{(1-\frac{\nu}{E}S_{xx}-\frac{\nu}{E}S_{yy})(1+\frac{\nu}{E}S_{xx}+\frac{\nu}{E}S_{yy})[1+\frac{\nu}{E}S_{yy}-\frac{\nu}{E}S_{xx}]}{b_s(1+\frac{\nu}{E}S_{yy}-\frac{\nu}{E}S_{xx})+2a_s(\frac{\nu}{E}S_{yy})} \right].$$

Or, $b_e a_s[1 - (\frac{\nu}{E}S_{xx})^2 - 2(\frac{\nu}{E}S_{xx})(\frac{\nu}{E}S_{yy}) - (\frac{\nu}{E}S_{yy})^2]$

$\quad + 2b_e(\frac{\nu}{E}S_{xx})[b_s(1 + \frac{S_{yy}\nu}{E} - \frac{S_{xx}\nu}{E}) + 2a_s(\frac{\nu}{E}S_{yy})]$

$= a_s b_s(1 - \frac{\nu}{E}S_{xx} - \frac{\nu}{E}S_{yy})(1 + \frac{\nu}{E}S_{xx} + \frac{\nu}{E}S_{yy})(1 + \frac{S_{yy}\nu}{E} - \frac{S_{xx}\nu}{E})$.

Or, $b_e a_s[1 - (\frac{\nu}{E}S_{xx})^2 - (\frac{\nu}{E}S_{yy})^2] - b_e a_s[2(\frac{\nu}{E}S_{xx})(\frac{\nu}{E}S_{yy})]$

$\quad + 2b_e(\frac{\nu}{E}S_{xx})[2a_s(\frac{\nu}{E}S_{yy})] + 2b_e(\frac{\nu}{E}S_{xx})[b_s(1 + \frac{S_{yy}\nu}{E} - \frac{S_{xx}\nu}{E})]$

$= a_s b_s(1 - \frac{\nu}{E}S_{xx} - \frac{\nu}{E}S_{yy})(1 + \frac{\nu}{E}S_{xx} + \frac{\nu}{E}S_{yy})(1 + \frac{S_{yy}\nu}{E} - \frac{S_{xx}\nu}{E})$.

Or, $b_e a_s[1 - (\frac{\nu}{E}S_{xx})^2 - (\frac{\nu}{E}S_{yy})^2] + 2b_e a_s(\frac{\nu}{E}S_{xx})(\frac{\nu}{E}S_{yy})$

$\quad + 2b_e(\frac{\nu}{E}S_{xx})[b_s(1 + \frac{S_{yy}\nu}{E} - \frac{S_{xx}\nu}{E})]$

$= a_s b_s(1 - \frac{\nu}{E}S_{xx} - \frac{\nu}{E}S_{yy})(1 + \frac{\nu}{E}S_{xx} + \frac{\nu}{E}S_{yy})(1 + \frac{S_{yy}\nu}{E} - \frac{S_{xx}\nu}{E})$.

Or, $b_e a_s[1 - (\frac{\nu}{E}S_{xx} - \frac{\nu}{E}S_{yy})^2]$

$\quad + 2b_e(\frac{\nu}{E}S_{xx})[b_s(1 + \frac{S_{yy}\nu}{E} - \frac{S_{xx}\nu}{E})]$

$= a_s b_s(1 - \frac{\nu}{E}S_{xx} - \frac{\nu}{E}S_{yy})(1 + \frac{\nu}{E}S_{xx} + \frac{\nu}{E}S_{yy})(1 + \frac{S_{yy}\nu}{E} - \frac{S_{xx}\nu}{E})$.

Or, $b_e a_s[1 - (\frac{S_{xx}\nu}{E} - \frac{S_{yy}\nu}{E})][1 + (\frac{\nu}{E}S_{xx} - \frac{\nu}{E}S_{yy})]$

$\quad + 2b_e(\frac{\nu}{E}S_{xx})[b_s(1 + \frac{S_{yy}\nu}{E} - \frac{S_{xx}\nu}{E})]$

$= a_s b_s(1 - \frac{\nu}{E}S_{xx} - \frac{\nu}{E}S_{yy})(1 + \frac{\nu}{E}S_{xx} + \frac{\nu}{E}S_{yy})(1 + \frac{S_{yy}\nu}{E} - \frac{S_{xx}\nu}{E})$.

Dividing both side by $(1 + \frac{S_{yy}\nu}{E} - \frac{S_{xx}\nu}{E})$, we obtain

$b_e a_s[1 + (\frac{\nu}{E}S_{xx} - \frac{\nu}{E}S_{yy})] + 2b_e(\frac{\nu}{E}S_{xx})[b_s]$

$= a_s b_s(1 - \frac{\nu}{E}S_{xx} - \frac{\nu}{E}S_{yy})(1 + \frac{\nu}{E}S_{xx} + \frac{\nu}{E}S_{yy})$.

Or,

$$b_e = \frac{a_s b_s(1 - \frac{\nu}{E}S_{xx} - \frac{\nu}{E}S_{yy})(1 + \frac{\nu}{E}S_{xx} + \frac{\nu}{E}S_{yy})}{a_s[1 + (\frac{\nu}{E}S_{xx} - \frac{\nu}{E}S_{yy})] + 2(\frac{\nu}{E}S_{xx})[b_s]}. \quad (8.2.3)$$

Furthermore, at the tip, using equation (8.2.2) and equation (8.2.3)

$$\sigma_{yy} = S_{yy}\left[1 + 2\frac{a_e}{b_e}\right] - S_{xx} = S_{yy} - S_{xx} + 2S_{yy}\left(\frac{a_e}{b_e}\right).$$

$$= S_{yy} - S_{xx}$$

$$+2S_{yy}\left[\frac{a_s b_s (1-\frac{\nu}{E}S_{xx}-\frac{\nu}{E}S_{yy})(1+\frac{\nu}{E}S_{xx}+\frac{\nu}{E}S_{yy})}{b_s(1+\frac{\nu}{E}S_{yy}-\frac{\nu}{E}S_{xx})+2a_s(\frac{\nu}{E}S_{yy})}\right]$$

$$\left[\frac{a_s[1+(\frac{\nu}{E}S_{xx}-\frac{\nu}{E}S_{yy})]+2(\frac{\nu}{E}S_{xx})[b_s]}{a_s b_s (1-\frac{\nu}{E}S_{xx}-\frac{\nu}{E}S_{yy})(1+\frac{\nu}{E}S_{xx}+\frac{\nu}{E}S_{yy})}\right].$$

Or, $\sigma_{yy} =$

$$S_{yy} - S_{xx} + 2S_{yy}\left[\frac{a_s(1 + \frac{\nu}{E}S_{xx} - \frac{\nu}{E}S_{yy}) + 2b_s(\frac{\nu}{E}S_{xx})}{b_s(1 + \frac{\nu}{E}S_{yy} - \frac{\nu}{E}S_{xx}) + 2a_s(\frac{\nu}{E}S_{yy})}\right].$$
$$(8.3.1)$$

Appendix F

Computing a_e and b_e - Green strain

When $a_s = 1$, $b_s = 0.1$, $\nu = 0.3$, $\frac{S_{xx}}{E} = 0$, and $\frac{S_{yy}}{E} = 0.1$,

using equation (10.1.1), $(b_e) = (b_s)\sqrt{1 + 2\nu(\frac{S_{yy}}{E})}$,

$(b_e) = 0.1\sqrt{1 + 2(0.3)(0.1)} = 0.1\sqrt{1.06} = 0.1029563014$.

Consequently, using equation (9.1.3), we can compute a_e

$$= -\left(2\frac{\nu}{E}S_{yy}\frac{a_s^2}{b_e}\right) + a_s\sqrt{\left(2\frac{\nu}{E}S_{yy}\frac{a_s}{b_e}\right)^2 + 1 + 2\left(\frac{\nu}{E}\right)(S_{xx} - S_{yy})}$$

$$= -\left(2(0.3)(0.1)\frac{1^2}{0.1\sqrt{1.06}}\right)$$

$$+ (1)\sqrt{\left(2(0.3)(0.1)\frac{1}{0.1\sqrt{1.06}}\right)^2 + 1 - 2\left((0.3)(0.1)\right)}$$

$$= \sqrt{[\tfrac{2(0.3)}{\sqrt{1.06}}]^2 + 0.94} - [\tfrac{2(0.3)}{\sqrt{1.06}}] = \sqrt{\tfrac{0.36}{1.06} + 0.94} - [\tfrac{0.6\sqrt{1.06}}{1.06}]$$

$$= \sqrt{0.3396.. + 0.94} - [\tfrac{0.6(1.029..)}{1.06}] = \sqrt{1.2796..} - [\tfrac{0.6177..}{1.06}]$$

$$= 1.1312040671.. - 0.5827715174.. = 0.5484325497..$$

And when $S_{yy} = 0$, equation (9.1.4) simplifies to

$$(b_e)^2 \frac{4(b_s)^2(\tfrac{\nu}{E})S_{xx}}{a_s(b_e)\sqrt{\left[1+2(\tfrac{\nu}{E})(S_{xx})\right]}} + (b_e)^2 - (b_s)^2\left[1 - 2(\tfrac{\nu}{E})S_{xx}\right] = 0.$$

Or, $(b_e)^2 + (b_e)\dfrac{4(b_s)^2(\tfrac{\nu}{E})S_{xx}}{a_s\sqrt{\left[1+2(\tfrac{\nu}{E})(S_{xx})\right]}} - (b_s)^2\left[1 - 2(\tfrac{\nu}{E})S_{xx}\right] = 0.$

Or, $(b_e) = -\dfrac{4(b_s)^2(\tfrac{\nu}{E})S_{xx}}{2a_s\sqrt{\left[1+2(\tfrac{\nu}{E})(S_{xx})\right]}}$

$$\pm\frac{1}{2}\sqrt{\frac{16(b_s)^4(\tfrac{\nu}{E})^2 S_{xx}^2}{a_s^2\left[1+2(\tfrac{\nu}{E})(S_{xx})\right]} + 4(b_s)^2\left[1 - 2(\tfrac{\nu}{E})S_{xx}\right]}.$$

Or, $(b_e) = -\dfrac{2(b_s)^2(\tfrac{\nu}{E})S_{xx}}{a_s\sqrt{\left[1+2\left(\tfrac{\nu}{E}\right)(S_{xx})\right]}}$ (discarding negative root)

$$+\frac{2b_s}{2}\sqrt{\frac{4(b_s)^2(\tfrac{\nu}{E})^2 S_{xx}^2}{a_s^2\left[1+2(\tfrac{\nu}{E})(S_{xx})\right]} + \left[1 - 2(\tfrac{\nu}{E})S_{xx}\right]},.$$

Or, $(b_e) = -\dfrac{2b_s(\tfrac{b_s}{a_s})(\tfrac{\nu}{E})S_{xx}}{\sqrt{\left[1+2(\tfrac{\nu}{E})(S_{xx})\right]}}$

$$+b_s\sqrt{\frac{4(\tfrac{b_s}{a_s})^2(\tfrac{\nu}{E})^2 S_{xx}^2 + \left[1 - 4(\tfrac{\nu}{E})^2 S_{xx}^2\right]}{\left[1+2(\tfrac{\nu}{E})(S_{xx})\right]}}.$$

Or,

$$b_e = \frac{b_s\sqrt{4(\frac{b_s}{a_s})^2(\frac{\nu}{E})^2 S_{xx}^2 + \left[1 - 4(\frac{\nu}{E})^2 S_{xx}^2\right]} - 2b_s(\frac{b_s}{a_s})(\frac{\nu}{E})S_{xx}}{\sqrt{\left[1 + 2(\frac{\nu}{E})(S_{xx})\right]}}.$$

(10.1.2)

For example, when $a_s = 1$, $b_s = 0.1$, $\nu = 0.3$, $\frac{S_{xx}}{E} = 0.1$, and $\frac{S_{yy}}{E} = 0$,

$$(b_e) = \frac{b_s\sqrt{4(\frac{b_s}{a_s})^2(\frac{\nu}{E})^2 S_{xx}^2 + \left[1 - 4(\frac{\nu}{E})^2 S_{xx}^2\right]} - 2b_s(\frac{b_s}{a_s})(\frac{\nu}{E})S_{xx}}{\sqrt{\left[1 + 2(\frac{\nu}{E})(S_{xx})\right]}}$$

$$= \frac{(0.1)\sqrt{4(0.1)^2(0.3)^2(0.1)^2 + \left[1 - 4(0.3)^2(0.1)^2\right]} - 2(0.1)(0.1)(0.3)(0.1)}{\sqrt{\left[1 + 2(0.3)(0.1)\right]}}$$

$$= \frac{(0.1)\sqrt{4(0.000009) + \left[1 - 4(0.0009)\right]} - 2(0.0003)}{\sqrt{\left[1 + 2(0.03)\right]}}$$

$$= \frac{\sqrt{0.000036 + \left[1 - 0.0036\right]} - 0.0006}{10\sqrt{\left[1 + 0.06\right]}} = \frac{\sqrt{0.996436} - 0.0006}{10\sqrt{[1.06]}}$$

$$= \frac{0.9982.. - 0.0006}{10[1.02956..]} = \frac{0.9976164094}{10.2956301409} = 0.0968970714.$$

Consequently, using equation (9.1.3), we can compute

a_e

$$= -\left(2\frac{\nu}{E}S_{yy}\frac{a_s^2}{b_e}\right) + a_s\sqrt{\left(2\frac{\nu}{E}S_{yy}\frac{a_s}{b_e}\right)^2 + 1 + 2\left(\frac{\nu}{E}\right)(S_{xx} - S_{yy})}$$

$$= -\left(0\right) + a_s\sqrt{\left(0\right)^2 + 1 + 2\left(\frac{\nu}{E}\right)(S_{xx} - 0)}$$

$$= a_s\sqrt{1 + 2\nu\left(\frac{S_{xx}}{E}\right)} = (1)\sqrt{1 + 2(0.3)\left(0.1\right)}$$

$$= \sqrt{1.06} = 1.02956301409.$$

However, when $a_s = 1$, $b_s = 0.1$, $\nu = 0.3$, $\frac{S_{xx}}{E} = 0.15$, and $\frac{S_{yy}}{E} = 0.1$,

$$(b_e)^2 \frac{4(b_s)^2(\frac{\nu}{E})S_{xx}}{-\left(2\frac{\nu}{E}S_{yy}a_s^2\right) + a_s\sqrt{\left(2\frac{\nu}{E}S_{yy}a_s\right)^2 + (b_e)^2\left[1 + 2\left(\frac{\nu}{E}\right)(S_{xx} - S_{yy})\right]}}$$

$$+ (b_e)^2 - (b_s)^2\left[1 + 2(\frac{\nu}{E})(S_{yy} - S_{xx})\right] = 0 \qquad (9.1.4)$$

simplifies to

$$\frac{(b_e)^2[4(0.1)^2(0.3)(0.15)]}{-\left(2(0.3)(0.1)(1)^2\right) + (1)\sqrt{\left(2(0.3)(0.1)(1)\right)^2 + (b_e)^2\left[1 + 2\left(0.3\right)(0.15 - 0.1)\right]}}$$

$$+ (b_e)^2 - (0.1)^2\left[1 + 2(0.3)(0.1 - 0.15)\right] = 0.$$

Or, $$\frac{(b_e)^2[4(0.01)(0.045)]}{-\left(2(0.03)\right) + \sqrt{\left(2(0.03)\right)^2 + (b_e)^2\left[1 + 2\left(0.3\right)(0.05)\right]}}$$

$$+ (b_e)^2 - (0.01)\left[1 - 2(0.3)(0.05)\right] = 0.$$

Or, $$\frac{(b_e)^2[0.0018]}{-\left(0.06\right) + \sqrt{\left(0.06\right)^2 + (b_e)^2\left[1 + (0.03)\right]}}$$

$$+(b_e)^2 - (0.01)\left[1 - 0.03\right] = 0.$$

Or,

$$\frac{(b_e)^2(0.0018)}{-0.06 + \sqrt{0.0036 + (b_e)^2 1.03}} + (b_e)^2 - 0.0097 = 0. \quad (F.0.1)$$

Using trial and error method described in section 10.1, we conclude that $b_e = 0.0969215694537$. Now, using equation (9.1.3), we can compute

a_e

$$= -\left(2\frac{\nu}{E}S_{yy}\frac{a_s^2}{b_e}\right) + a_s\sqrt{\left(2\frac{\nu}{E}S_{yy}\frac{a_s}{b_e}\right)^2 + 1 + 2\left(\frac{\nu}{E}\right)(S_{xx} - S_{yy})}$$

$$= -\left(2(0.3)(0.1)\frac{1^2}{0.0969..}\right)$$

$$+ (1)\sqrt{\left(2(0.3)(0.1)\frac{1}{0.0969..}\right)^2 + 1 + 2\left(0.3\right)(0.15 - 0.1)}$$

$$= -\left(\frac{0.06}{0.0969..}\right) + \sqrt{\left(\frac{0.06}{0.0969..}\right)^2 + 1 + (0.6)(0.05)}$$

$$= \frac{1}{0.0969..}\sqrt{\left(0.06\right)^2 + (1.03)(0.0969..)^2} - \left(\frac{0.06}{0.0969..}\right)$$

$$= \frac{\sqrt{0.0036 + 0.0096756043}}{0.0969..} - \left(\frac{0.06}{0.0969..}\right) = \frac{\sqrt{0.0132756043}}{0.0969..} - \frac{0.06}{0.0969..}$$

$$= \frac{0.1152198087}{0.0969..} - \frac{0.06}{0.0969..} = \frac{0.0552198087}{0.0969..} = 0.5697370473$$

To compare these results with the use of engineering strain we recall equation (8.2.2) and equation (8.2.3). Therefore,

$$a_e = a_s b_s \left[\frac{[1-(\frac{\nu}{E}S_{xx})-(\frac{\nu}{E}S_{yy})][1+(\frac{\nu}{E}S_{xx})+(\frac{\nu}{E}S_{yy})]}{b_s(1+\frac{\nu}{E}S_{yy}-\frac{\nu}{E}S_{xx})+2a_s(\frac{\nu}{E}S_{yy})} \right]$$

$$= (1)(0.1) \left[\frac{[1-(0.3)(0.15)-(0.3)(0.1)][1+(0.3)(0.15)+(0.3)(0.1)]}{(0.1)(1+(0.3)(0.1)-(0.3)(0.15))+2(1)(0.3)(0.1)} \right]$$

$$= (0.1) \left[\frac{[1-(0.045)-(0.03)][1+(0.045)+(0.03)]}{(0.1)(1+(0.03)-(0.045))+2(0.03)} \right]$$

$$= \left[\frac{(0.1)[1-0.075][1+0.075]}{(0.1)(1+0.03-0.045)+0.06} \right] = \left[\frac{(0.1)[0.925][1.075]}{(0.1)(0.985)+0.06} \right]$$

$$= \left[\frac{[0.0925][1.075]}{0.0985+0.06} \right] = \left[\frac{0.0994375}{0.1585} \right] = 0.6273659306.$$

And,

$$b_e = \frac{a_s b_s (1-\frac{\nu}{E}S_{xx}-\frac{\nu}{E}S_{yy})(1+\frac{\nu}{E}S_{xx}+\frac{\nu}{E}S_{yy})}{a_s[1+(\frac{\nu}{E}S_{xx}-\frac{\nu}{E}S_{yy})]+2(\frac{\nu}{E}S_{xx})[b_s]}$$

$$= \frac{(1)(0.1)(1-(0.3)(0.15)-(0.3)(0.1))(1+(0.3)(0.15)+(0.3)(0.1))}{(1)[1+((0.3)(0.15)-(0.3)(0.1))]+2((0.3)(0.15))[0.1]}$$

$$= \frac{(0.1)(1-(0.045)-(0.03))(1+(0.045)+(0.03))}{[1+((0.045)-(0.03))]+2(0.045)[0.1]}$$

$$= \frac{(0.1)(1-0.075)(1+0.075)}{[1+0.015]+(0.09)[0.1]} = \frac{(0.1)(0.925)(1.075)}{[1.015]+(0.009)}$$

$$= \frac{(0.0925)(1.075)}{1.024} = \frac{0.0994375}{1.024} = 0.0971069335.$$

However, when $a_s = 1$, $b_s = 0.1$, $\nu = 0.3$, $\frac{S_{xx}}{E} = -0.15$, and $\frac{S_{yy}}{E} = -0.1$,

$$(b_e)^2 \frac{4(b_s)^2(\frac{\nu}{E})S_{xx}}{-\left(2\frac{\nu}{E}S_{yy}a_s^2\right)+a_s\sqrt{\left(2\frac{\nu}{E}S_{yy}a_s\right)^2+(b_e)^2\left[1+2\left(\frac{\nu}{E}\right)(S_{xx}-S_{yy})\right]}}$$

$$+ (b_e)^2 - (b_s)^2 \left[1+2(\frac{\nu}{E})(S_{yy}-S_{xx})\right] = 0 \qquad (9.1.4)$$

simplifies to

$$\frac{(b_e)^2[4(0.1)^2(0.3)(-0.15)]}{-\left[2(0.3)(-0.1)(1)^2\right]+(1)\sqrt{\left[2(0.3)(-0.1)(1)\right]^2+(b_e)^2\left[1+2(0.3)(-0.15+0.1)\right]}}$$

$$+(b_e)^2-(0.1)^2\left[1+2(0.3)(-0.1+0.15)\right]=0.$$

Or,
$$\frac{(b_e)^2[4(0.01)(-0.045)]}{-\left(2(-0.03)\right)+\sqrt{\left(2(-0.03)\right)^2+(b_e)^2\left[1+2\left(0.3\right)(-0.05)\right]}}$$

$$+(b_e)^2-(0.01)\left[1+2(0.3)(0.05)\right]=0.$$

Or,
$$\frac{(b_e)^2[-0.0018]}{-\left(-0.06\right)+\sqrt{\left(-0.06\right)^2+(b_e)^2\left[1+(-0.03)\right]}}$$

$$+(b_e)^2-(0.01)\left[1+0.03\right]=0.$$

Or,

$$\frac{(b_e)^2(-0.0018)}{0.06+\sqrt{0.0036+(b_e)^2 0.97}}+(b_e)^2-0.0103=0. \quad \text{(F.0.2)}$$

Using trial and error method described in section 10.1, we conclude that $b_e = 0.10219819$. Now, using equation (9.1.3), we can compute

$$a_e$$

$$=-\left(2\frac{\nu}{E}S_{yy}\frac{a_s^2}{b_e}\right)+a_s\sqrt{(2\frac{\nu}{E}S_{yy}\frac{a_s}{b_e})^2+1+2(\frac{\nu}{E})(S_{xx}-S_{yy})}$$

$$=-\left(2(0.3)(-0.1)\frac{1^2}{0.1021..}\right)$$

$$+(1)\sqrt{\left(2(0.3)(-0.1)\frac{1}{0.1021..}\right)^2+1+2(0.3)(-0.15+0.1)}$$

$$= -\left(\frac{-0.06}{0.1021..}\right) + \sqrt{\left(\frac{-0.06}{0.1021..}\right)^2 + 1 + (0.6)(-0.05)}$$

$$= \frac{1}{0.1021..}\sqrt{\left(-0.06\right)^2 + (0.97)(0.1021..)^2} + \left(\frac{0.06}{0.1021..}\right)$$

$$= \frac{\sqrt{0.0036+0.0101311359}}{0.1021..} + \left(\frac{0.06}{0.1021..}\right) = \frac{\sqrt{0.0137311359}}{0.1021..} + \frac{0.06}{0.1021..}$$

$$= \frac{0.1171799297}{0.1021..} + \frac{0.06}{0.1021..} = \frac{0.1771799297}{0.1021..} = 1.7336895082$$

To compare these results with the use of engineering strain we recall equation (8.2.2) and equation (8.2.3). Therefore,

$$a_e = a_s b_s \left[\frac{[1-(\frac{\nu}{E}S_{xx})-(\frac{\nu}{E}S_{yy})][1+(\frac{\nu}{E}S_{xx})+(\frac{\nu}{E}S_{yy})]}{b_s(1+\frac{\nu}{E}S_{yy}-\frac{\nu}{E}S_{xx})+2a_s(\frac{\nu}{E}S_{yy})}\right]$$

$$= (1)(0.1)\left[\frac{[1-(0.3)(-0.15)-(0.3)(-0.1)][1+(0.3)(-0.15)+(0.3)(-0.1)]}{(0.1)(1+(0.3)(-0.1)-(0.3)(-0.15))+2(1)(0.3)(-0.1)}\right]$$

$$= (0.1)\left[\frac{[1+(0.045)+(0.03)][1-(0.045)-(0.03)]}{(0.1)(1-(0.03)+(0.045))+2(-0.03)}\right]$$

$$= \left[\frac{(0.1)[1+0.075][1-0.075]}{(0.1)(1-0.03+0.045)-0.06}\right] = \left[\frac{(0.1)[1.075][0.925]}{(0.1)(1.015)-0.06}\right]$$

$$= \left[\frac{[0.0925][1.075]}{0.1015-0.06}\right] = \left[\frac{0.0994375}{0.0415}\right] = 2.3960843373.$$

And,

$$b_e = \frac{a_s b_s (1-\frac{\nu}{E}S_{xx}-\frac{\nu}{E}S_{yy})(1+\frac{\nu}{E}S_{xx}+\frac{\nu}{E}S_{yy})}{a_s[1+(\frac{\nu}{E}S_{xx}-\frac{\nu}{E}S_{yy})]+2(\frac{\nu}{E}S_{xx})[b_s]}$$

$$= \frac{(1)(0.1)(1-(0.3)(-0.15)-(0.3)(-0.1))(1+(0.3)(-0.15)+(0.3)(-0.1))}{(1)[1+((0.3)(-0.15)-(0.3)(-0.1))]+2((0.3)(-0.15))[0.1]}$$

$$= \frac{(0.1)(1+(0.045)+(0.03))(1-(0.045)-(0.03))}{[1+((-0.045)-(-0.03))]+2(-0.045)[0.1]} = \frac{(0.1)(1+0.075)(1-0.075)}{[1-0.015]-(0.09)[0.1]}$$

$$= \frac{(0.1)(0.925)(1.075)}{[0.985]-(0.009)} = \frac{(0.0925)(1.075)}{0.976} = \frac{0.0994375}{0.976}$$

$$= 0.1018826844.$$

Appendix G

Computing a_e and b_e - Logarithmic strain

We need to solve equations (10.2.1) and (10.2.2) each containing two unknowns, a_e and b_e. We employ the trial and error method. We set up a 5x5 matrix in Excel (cells C4 to G8) where rows (4 to 8) represent values of a_e, columns (C to G) represent values of b_e, and individual cells represent corresponding values of the left side of equation (10.2.1).

First, we select 0.08, 0.09, 0.1, 0.11, and 0.12 as values of b_e since we expect b_e value for the solution to be within this range. We place these five values in the top row of the Table G.0.1 (above the 5x5 matrix in cells C3 to G3), as shown. Then in the first column (left of the 5x5 matrix in cells B4 to B8) under the heading ($a_e \downarrow b_e \rightarrow$) (in cell B3), we place 1 in each of the five cells (B4 to B8). We reserve rows 1 and 2 as well as column A for titles and comments.

Next, for the first selected b_e value (i.e. 0.08), using the method described in section 10.1, we find the value of $a_e(= 0.63194249785)$ that makes the left side of equation (10.2.1) zero (up to the desired accuracy, we use ten decimal places). We use this value of $a_e(= 0.63194249785)$ to replace 1 in the first cell under the heading $(a_e \downarrow b_e \rightarrow)$ i.e. cell B4.

We convert the left side of the equation (10.2.1) to its Excel format [i.e. EXP(((0.06 *B4)/C3)-0.015) - (1/B4)]. We place this Excel format (with = sign in front) in the second cell below $b_e = 0.08$ (i.e. cell C4) which is just right of the cell containing 0.63194249785 so that the cell (in the first row and the first column of the 5x5 matrix) shows the computed value as 0.000000000. We copy this Excel equation and paste it in the remaining twenty four cells of the 5x5 matrix. We ensure that the Excel equation in each cell involves the corresponding cell references for a_e and b_e by making necessary changes in each Excel equation.

At this stage, the first row corresponding to $a_e = 0.63194249785$ shows five more values i.e. 0.000000000, $-0.081177104, -0.143110058, -0.191877246$, and -0.231251963 under the five column headings $b_e = 0.08, 0.09,$ 0.1, 0.11, and 0.12. Accordingly, $a_e = 0.63194249785,$ $b_e = 0.08$ is a solution to equation (10.2.1). The second row corresponding to $a_e = 1.0000000000$ shows five more values i.e. $1.0854819925, 0.9187360590, 0.7949909856, 0.6997047261,$ and 0.6241750088 under the five column headings $b_e = 0.08, 0.09, 0.1, 0.11,$ and 0.12. The next three rows show the same values as the second row.

$a_e\downarrow$ $b_e\rightarrow$	0.08	0.09	0.1	0.11	0.12
0.63194249785	0.000000000	-0.081177104	-0.143110058	-0.191877246	-0.231251963
0.65566285167	0.085651977	0.000000000	-0.065230638	-0.116520602	-0.157883079
0.67646286824	0.157874273	0.068192772	0.000000000	-0.053551410	-0.096692680
0.69487529147	0.219795731	0.126463194	0.055592338	0.000000000	-0.044744198
0.71130577020	0.273606621	0.176948162	0.103642376	0.046197349	0.000000000

Table G.0.1: Values of the left side of equation (10.2.1) for selected five values of a_e and selected five values of b_e.

This Table shows five points which satisfy equation (10.2.1).
They are (0.63194249785, 0.08), (0.65566285167, 0.09), (0.6764286824, 0.1),
(0.69487529147, 0.11), and (0.71130577020, 0.12).

Now, we find another solution to equation (10.2.1). For this purpose, we change 1 in the second cell under the heading a_e (i.e. in cell B5) to a value that makes the second cell under the heading 0.09 (i.e. the cell in the second row and the second column of the 5x5 matrix, cell D5) equal to 0.000000000. By trial and error method, we find this value to be 0.6556628517. Similarly, we find values for the third, fourth, and fifth cells under the heading a_e to be 0.6764628682, 0.6948752915, and 0.7113057702 which make the third cell under the heading 0.1 (cell E6), the fourth cell under the heading 0.11 (cell F7), and the fifth cell under the heading 0.12 (cell G8), all equal to 0.000000000. The resulting matrix is shown in Table G.0.1.

We have now found five solutions to equation (10.2.1) which can be used to plot part of the curve representing equation (10.2.1). These solutions are listed under Table G.0.1.

We use the same procedure to find five solutions to equation (10.2.2) which can be used to plot part of the curve representing equation (10.2.2). The only difference in this procedure is that we select 0.64, 0.65, 0.66, 0.67, and 0.68 as values of a_e and find corresponding b_e values that satisfy equation (10.2.2). The resulting matrix is shown in Table G.0.2.

We have now found five solutions to equation (10.2.2) which can be used to plot part of the curve representing equation (10.2.2). These solutions are listed under Table G.0.2.

These ten points are plotted in Figure 10.2.1.

$a_e \downarrow$ $b_e \rightarrow$	0.097174183	0.097194343	0.097213904	0.097232892	0.097251333
0.64	0.000000000	0.000216369	0.000426225	0.000629857	0.000827539
0.65	-0.000216323	0.000000000	0.000209812	0.000413401	0.000611041
0.66	-0.000426048	-0.000209769	0.000000000	0.000203548	0.000401148
0.67	-0.000629472	-0.000413235	-0.000203508	0.000000000	0.000197561
0.68	-0.000826874	-0.000610679	-0.000400991	-0.000197523	0.0000000000

Table G.0.2: Values of the left side of equation (10.2.2) for selected five values of a_e and selected five values of b_e.

This Table shows five points which satisfy equation (10.2.2).
They are (0.64, 0.097174183), (0.65, 0.097194343), (0.66, 0.097213904), (0.67, 0.097232892), and (0.68, 0.097251333).

CORRECTIONS

I have found two typographical errors in the first book (titled *Understanding the Elastic Stress Field Around an Elliptical Hole in a Thin Plate (in the deformed configuration) ISBN 978-1-48359-262-6*).

    1) page 79, when $b_e/a_e = 0.2$, α_e should be 0.20273255 not 0.020273255.

    2) page 150, the first factor in the first term in the second line should be S_{xx} not S_{yy}.

A reviewer found one more error in the first book.

    3) page 91, the value for $\frac{b_e}{a_e}$ should be 0.024 not 0 : 024.

REVIEWS AND ENDORSEMENTS

A REVIEW

by

Amit Singh

Assistant Professor

Department of Mechanical Engineering

Indian Institute of Technology, Bombay, India

July 8, 2019

This monograph by Rajen Merchant is a welcome addition to the vast literature around the stress distribution in a thin plate with circular and elliptical holes. Kirsch [1] in 1898 came up with the first analytical solutions for the stress concentration around a circular hole in an infinite plate using Airy function approach. Then Inglis [2] in 1913 used complex potential functions and developed solutions for the stress field around elliptical holes in an infinite plate with the help of Elliptical Co-ordinate System (ECS). This led to further growth in literature [3, 4] where many such problems have been solved.

The attempts have also been made to develop closed-form solutions in Cartesian co-ordinates since solutions in Elliptical co-ordinates are not convenient for application to strength analysis [5]. However, any information about the deformed configuration is not explicit in these relations developed in Cartesian co-ordinates, moreover, the relations become cumbersome, therefore, the author likes the ECS and presented the stress field in terms of a shape parameter α_e of the deformed configuration in a monograph published earlier [6].

The present monograph develops relations for displacements and deformation gradients in ECS and uses four different definitions of strain (Engineering, Green, logarithmic and Almansi) as well as Hooke's law as constitutive equation, to obtain lengths of semi major and minor axes, a_e and b_e, respectively, of the deformed elliptical hole. Then, the monograph derives expressions for stress distribution in terms of material parameters, Young's Modulus E and Poisson's Ratio ν, applied stresses and initial geometry described by a_s and b_s.

The real substance of the whole book has already been presented by the author in the final Chapter 12 where summaries and conclusions of each chapter from 1 to 11 have been provided. The first two chapters are very interesting from the point of view of understanding nuances of the ECS when compared to Cartesian Co-ordinate System (CCS), especially for the beginner. The first chapter tries to find analogies of straight lines $x =$ constant and $y =$ constant in the CCS with ellipses $\alpha =$ constant and hyperbolas β $=$ constant, when the co-ordinate transformation of a point (x, y) in the CCS to (α, β) in the ECS reveals that $x = c \cosh\alpha \cos\beta$ and $y = c \sinh\alpha \sin\beta$, where $\tanh\alpha = b/a$, $a = c \cosh\alpha$ and $b = c \sinh\alpha$. Thereafter, the effects of changes in c and α upon the nature of elliptical and hyperbolic curves have been explored. Chapter 2 deals with the displacements in the ECS and concludes that displacements at the tip and top of the elliptical curve are horizontal and vertical, respectively, whereas they are not perpendicular to the ellipse at other points on the curve. Furthermore, it has also been shown that any displacement vector can be decomposed into two components, one due to change in c and the other due to change in α from the starting elliptical configuration to the end elliptical configuration. The analogy of a deformation with constant α and changing c with constant passenger co-ordinates with respect to driver in a moving car is illuminating and helps build the intuition among the readers about the meaning of α, β, and c in the ECS.

Chapter 3 provides a detailed derivation of both α in terms of x, y, c and x, y, β, and β in terms of x, y, c and x, y, α. Chapters 4 and 5 help derive the displacement and deformation gradients for the displacement $u \equiv (u_x, u_y)$ of a point from (x_s, y_s) to (x_e, y_e). Chapter 6 presents Hooke's law for the plane strain/stress and derives strains at the tip and top of the hole in terms of material constants, applied stresses and the ratio $\frac{a_e}{b_e}$. Chapter 7 presents four different definitions of strain. Chapter 8 provides the expressions for a_e and b_e for the deformed configuration when the engineering strain is chosen for Hooke's law showing relationship between the Cauchy stress and the engineering strain. This helps find solutions for stresses in final configuration and both tip and top of the holes have been analyzed in great details. The expressions for stresses were compared with Singh, Glinka and Dubey (1994) [7] and it was found that the results obtained by them are the special cases ($\nu = 1$) of the general solution developed in the monograph. A comparison with small deformation theory has also been provided and it was concluded that stress values obtained by small deformation theory serve as upper bound for the general results.

Chapter 9 develops equations for obtaining a_e and b_e when Green, logarithmic and Almansi strains are used in the analysis. Chapter 10 describes procedures for solving the equations obtained in Chapter 9 for Hooke's law based five different constitutive equations relating (a) the Green strain with the Cauchy stress, (b) the logarithmic strain with the Cauchy stress, (c) the logarithmic strain with the Kirchoff stress, (d) the Almansi strain with the Cauchy stress, and (e) the Almansi strain with the Kirchoff stress. Thus we get six different models, five described in Chapter 10 and one in Chapter 8.

Finally, Chapter 11 compares results for six different models under four loading conditions. It was found that, for the loading case $S_{xx} = 0$, b_e does not depend upon a_s for all six models, i.e., it does not matter if the initial hole is circular, elliptical or a sharp long crack. Similarly, when $S_{yy} = 0$, a_e does not depend upon b_s for all six models. The displacements of points on the contour of the elliptical hole were also compared for all four loading conditions and important conclusions were drawn with regard to expansion and contraction of the semi major and minor axes. Zero stress points and tip stress were also compared. In the last section of the Chapter, limitations of models were considered, where it was first noted that material non-linearity was not discussed at all and secondly, the geometric non-linearity implicit in strain definitions may not find any solutions for particular loading conditions and/or initial geometry.

Overall, the chapters have been written clearly with enough details and therefore can aid teaching of any undergraduate or graduate level courses without much difficulties. The uncluttered graphs and explicit tables help understand the equations and procedures. There are couple of suggestions though:

- The notion of β appears on page 6 without any introduction. It assumes familiarity with the ECS or the monograph published earlier. We felt that upper portion of the first paragraph on page 7 can be shifted to the beginning of second paragraph of page 6.

- The trial and error procedures introduced in Chapter 10 are not very sophisticated. Some advance methods to solve the non-linear equations of Chapter 9 based on iteration and convergence in different function spaces such as non-linear Gauss - Seidel method and other methods [8, 9] could have been introduced.

Regardless of these minor issues, which can be easily addressed, the monograph is excellent in convincing the reader that working with the ECS and final shape parameters in deformed configuration makes the analytical solutions simpler to understand, which also helps think about the problems in a much clearer way with better intuitions. Its inclusion for both students and teachers of the subject as a nice text cannot be recommended more.

References

(1) G. Kirsch. *V.D.I.*, 47:797807, 1898.

(2) C. Inglis. Stresses in a plate due to the presence of cracks and sharp corners. *Transactions of the Institute of Naval Architects*, 55:219241, 1913.

(3) S. Timoshenko and J.N. Goodier. Theory of Elasticity. *Engineering Societies Monographs. McGraw-Hill*, 1951.

(4) W.D. Pilkey. Peterson's Stress Concentration Factors. *Wiley*, 1997. ISBN 9780471538493.

(5) T. Kanezaki, K. Nagata, and Y. Murakami. New Closed-form Solution by Cartesian Co-ordinate for Stress Distribution Around Elliptic Hole and Its Applications. *Journal of Solid Mechanics and Materials Engineering*, 1(2):232243, 2007.

(6) Rajen Merchant. Understanding the Elastic Stress Field Around an Elliptical Hole in a Thin Plate (in the deformed configuration). *Bookbaby*, 2017. ISBN 9781483592626.

(7) M. Singh, G. Glinka, and R. Dubey. Notch and Crack Analysis as a Moving Boundary Problem. *Engineering Fracture Mechanics*, 47(4):479492, 1994.

(8) W. H. Press et al. Numerical Recipes in C++: The Art of Scientific Computing. *Cambridge University Press*, 2nd edition, 2002.

(9) C. Grosan and A. Abraham. A New Approach for Solving Non-linear Equations Systems. *IEEE Transactions on Systems, Man, and Cybernetics PART A: Systems and Humans*, 38(3):698714, 2008.

Author's note : The first suggestion by Prof. Amit Singh is implemented by reshuffling the text on pages 6 and 7. For the second suggestion, a paragraph on top of Chapter 10 has been inserted noting that a discussion on sophisticated computational procedures is beyond the scope of this book. However, if there is sufficient interest, such a discussion in detail may be presented in future as part of the third book on "Applications" where analytical results can be compared with results from FEM codes and/or experimental studies. Suggestion have been made to use analytical results as benchmarks to evaluate FEM codes.

AN ENDORSEMENT

by

Rashmi C. Desai

Professor Emeritus, Department of Physics

University of Toronto, Toronto, Canada

July 12, 2019

The science of elasticity and especially the stress-strain relations are fundamentally important due to their relevance both in Physics and in a large variety of applications. Non-linear relations play very important role in analysis of stresses and associated strains. The first two books by Rajen Merchant in this area are remarkable and outstanding. First one deals with geometric non-linearity and develops a stress function for elastic stress for a thin plate system with an elliptical hole. The second book describes his derivations and ideas about non-linear strain displacement expressions for the same system.

Rajen uses the Elliptical co-ordinate system since it is much more versatile. This is very challenging; his derivations are impressive and resulting exact expressions are truly new results emerging from his research. He has compared his theory and various approximate models by others, which is also very useful to researchers in this area. This series of books by Rajen Merchant should be in various University libraries, so that they are available to students in Civil Engineering, Physics and other areas where elasticity theory is of use and value. I congratulate Rajen for this novel accomplishment.

AN ENDORSEMENT
by
Amin Eshraghi Ph.D., P.Eng.,
Senior Materials and Structures Engineer

Ottawa, Canada

July 18, 2019

In his first book, Rajen derived closed form solution for the stress field for an elliptical hole in an infinite plate using the stress equilibrium equations. The stress components were expressed in terms of a single parameter α_e. The present book develops expressions for deformation gradients in the Elliptical co-ordinate system. These deformation gradients are used to set up both non-linear and linear strain-displacement relations and are used to determine the parameters a_e, semi-major axis and b_e, semi-minor axis of the deformed elliptical hole for linear elastic materials.

These two books provide detailed analysis for the stress and the strain fields of cracks and holes in linear elastic thin plates. The developed expressions in these two books can be used for a wide variety of applications. I commend Rajen for these well written books and I am sure these two books will be strong references for courses where these topics are taught.

AN ENDORSEMENT
by
Anup Sahoo Ph.D.,
Senior Engineer/Scientist
Toronto, Canada
July 24, 2019

Rajen's first book dealt primarily with geometric non-linearity. This second book deals primarily with non-linearity in strain-displacement relations and non-linearity in definitions of strain. Six different models based on four definitions of strain and two definitions of stress are developed. It is shown that all expressions for all models contain E, Young's Modulus and ν, Poisson's Ratio. Researchers around the world will find these two books very useful while analyzing problems that involve thin sheets made of linear elastic materials. Some libraries have already included the first book in their collection.

Index